Creating a Kaizen Culture

About the Authors

Jon Miller has 20 years of experience in the field of *kaizen*. He is the CEO of the Kaizen Institute Consulting Group. Born in Japan and living there for 18 years, he is fluent in its language and culture, allowing him to work closely with Japanese *kaizen* gurus. He was co-founder and CEO of Gemba Research, at which he spent 15 years as a senior consultant and trainer, growing the firm to over 40 people before merging it with the Kaizen Institute in 2011. Jon has traveled to more than 30 countries and has wide experience designing global Lean deployment programs for many organizations. He is the author of numerous articles, and translated and edited *Taiichi Ohno's Workplace Management* as well as contributing to other McGraw-Hill Professional books, including *Gemba Kaizen*, Second Edition, and *Kaizen in Logistics and Supply Chains*.

Mike Wroblewski has more than 25 years of experience in industrial engineering, manufacturing management, and *kaizen*. He is a Director of the Kaizen Institute USA. He has also served as an internal consultant and Lean *Sensei* with Batesville Casket Company, with three factories winning the *IndustryWeek* "Top 10 Best Plants" award a total of nine times and one facility awarded *Assembly Magazine*'s North American Plant of the Year. Mike is a Certified Six Sigma Black Belt from the Milwaukee School of Engineering. He is a popular keynote speaker at national conferences, sharing his insights and experiences leading *kaizen*.

Jaime Villafuerte has 20 years of experience in the strategic and tactical deployment of continuous improvement. He is a Lean Six Sigma Director at Jabil, Inc., where he leads the development and deployment of Lean Six Sigma transformation, reaching 165,000+ employees in 33 countries. This includes designing learning solutions, developing the competencies of *kaizen* regional and site Lean managers, and authoring new training courses. Jaime is an ASQ Certified Six Sigma Master Black Belt and Lean Gold Certified by SME, AME, and the Shingo Prize. He served as a Baldrige National Award Examiner in 2010. He is the author of *The Lean 6 Sigma Deployment Memory Jogger*.

Creating a Kaizen Culture

Align the Organization,
Achieve Breakthrough Results,
and Sustain the Gains

Jon Miller

Mike Wroblewski

Jaime Villafuerte

New York   Chicago   San Francisco
Athens   London   Madrid
Mexico City   Milan   New Delhi
Singapore   Sydney   Toronto

ISBN   978-0-07-182685-3
MHID   0-07-182685-8

The pages within this book were printed on acid-free paper.

Sponsoring Editor Judy Bass	**Copy Editor** James K. Madru
Editing Supervisor Stephen M. Smith	**Proofreader** Claire Splan
Production Supervisor Pamela A. Pelton	**Indexer** Judy Davis
Acquisitions Coordinator Amy Stonebraker	**Art Director, Cover** Jeff Weeks
Project Manager Patricia Wallenburg, TypeWriting	**Composition** TypeWriting

CONTENTS

FOREWORD

My personal journey with *kaizen* culture began when I first became chief medical officer at ThedaCare. My big concern was that we seemed to be unable to deliver reliable care to our patients. I had read the standard management books, but the problem that really frustrated me was that while we could make improvements, we couldn't sustain them. The metaphor I like to use is we'd shine the flashlight into one corner and improve that area, but as soon as we took the flashlight and shone it in the other corner, the results went right back to baseline.

I had visited a number of healthcare facilities, and although some were better than others, they weren't much better. I was searching for a method that could deliver zero defects. Was there an operating system or a process that could give us the consistent quality that manufacturers had achieved in the auto and aerospace industries? Could we apply those principles to healthcare to get to zero defects?

These questions led me to lean manufacturing and to study Toyota, Boeing, and finally Ariens, where we spent a day watching the workers build a snowblower. That's when I realized it's about the culture. The staff on the snowblower floor were working together in teams, and identifying problems and solving them. They didn't have some manager telling them what to do. They simply were in charge of managing the quality of that product. And they were able to be in charge because they had the knowledge and tools to identify and solve problems. That was a culture that didn't exist in our healthcare organization.

I made the leap of faith that working with patients isn't really so different from working with snowblowers—if you add caring and compassion, it's still a process. And they were probably delivering better care to those snowblowers than we were to our patients. It was a real slap in the face to see that we had all these highly intelligent sophisticated people, and they couldn't work together to actually identify and remove defects.

In hindsight, this shouldn't have been surprising. Most administrators are still back in the 1930s with Sloan-style management, which is the top-down autocracy taught in most business schools today. It's ingrained in our training, especially in healthcare, that if you make a mistake, you're a bad person. So we wind up with a shame and blame culture where people hide their mistakes and nothing gets improved.

In reality, when you actually study defects—I don't care in what industry—it's almost always a process problem rather than a people problem. We have the most highly trained people on earth in healthcare, so for us to think mistakes are due to bad people is just ludicrous. It also surprises me that our profession, which is based on science, doesn't utilize the plan-do-study-act (PDSA) method for problem solving.

To change the culture leaders must make it clear that if a team member does make a medication error, it's a process problem, not a people problem. Our job as managers and leaders is to redesign the process that delivered the defect. The culture changes when the leaders focus on the process and quit blaming the people.

Continuous improvement, however, can't just evolve from a project or series of projects. I've visited 125 health systems in 12 countries in the last 8 years, and the piece that's missing is the management system. Many organizations utilize lean tools, and I've seen good adaptation of human development and HR processes, but I rarely see application of the lean management system. So teaching the lean management system has been my main focus when I work with other organizations.

If we expect workers to act differently, we can no longer practice Sloan-style autocratic management—it simply won't work. Cultural transformation has really got to occur at the manager and leader level, and it can't be delegated. The leaders must lead the transformation, not just say "oh well, let's just do some *kaizen* and we'll be all set with the workers." The leadership level is where most organizations are failing at the lean transformation.

I think lean can be no less than a transformation to a culture of continuous improvement from top to bottom. It should infect the CEO, the front-line administrative assistant, and everybody in between with a virus of identifying and solving problems at the front line. This takes an enormous effort in terms of change management and personal commitment. Every leader has to come to terms with the notion that "change is great as long as I don't have to."

Kaizen culture doesn't happen overnight. In our case, there was a lot of standard work to create first, and it was 6 or 7 years before managers began to say, "improvement is my job." But today, ThedaCare is famous, and hosts many visitors—750 CEOs and senior executives this year alone. Visitors will go to huddles where staff members spend 15 minutes twice a day at our visual tracking centers throughout the hospitals and clinics, identifying problems they've encountered during their shift. For each problem, a team member will step up to take ownership. If there are barriers, it's the manager's role to support the team member by removing that barrier. But it is the team member who will solve the problem.

My experience, and that of others I work with in North America and the world, suggests that if you really want to get deep employee engagement in the work, lean is the way to accomplish that. However, we still have a major learning gap not only in healthcare, but in any organization, any industry at this point.

The book you are about to read addresses many of the issues I have briefly summarized above. The real-world experience of the authors is obvious as they describe problems and solutions that they and others have used to build a *kaizen* culture. As the authors point out, culture is determined by a set of behaviors that start at the level of the leader and filter through the rest of the organization. This book will give hope that you can change your organization too. But remember, the time to get started is now, not next week, next month, or next year. The system is broken, you are in charge, and *kaizen* culture is the solution.

John Toussaint, MD
Founder and CEO
ThedaCare Center for Healthcare Value

INTRODUCTION

––––––––––

What Problem Are We Solving?

There is no longer any question as to the effectiveness of *kaizen*. In fact, team-based scientific problem solving is more important today than ever before in building peace and prosperity in our interconnected and rapidly changing world. This book is not intended as an authoritative scientific study on the impact of organizational culture on performance, although the renewal of such study is long overdue. Rather, it is the beginning of a conversation about what it means for organizations to excel sustainably over long periods of time despite the rapid and often drastic changes around us. It is our hypothesis that organizations where we can see evidence of *kaizen* cultures are the ones that will be able to experience lasting success.

These *kaizen* cultures are based on putting into action a set of core beliefs, including but not limited to engaging the total workforce, servant leadership, visualization of the real condition of things, respect for people, appreciation for standards, scientific problem solving, alignment of purpose not only with customers but also with broader stakeholders, curiosity, humility, and a view to the long term. The examples we will introduce in this book are such organizations or those on the way to becoming *kaizen* cultures. We have not encountered, nor have we spent effort in searching for, organizations that succeed over the long term yet possess what we may call anti-*kaizen* cultures. It is theoretically possible that arrogant, short-term-focused, close-minded, disrespectful-of-people, resisting-change, and self-serving bureaucratic organizations succeed over long periods of time. However, we believe such a result is unlikely in a world where we have choices. Customers have choices, and in an increasingly connected world, customers learn about the toxic cultures of companies and choose to buy from others. People choose not to work for these organizations, and as a result such organizations are starved of talent. Inevitably, even financial institutions and traders decide that there are better places to put their money. Even if an entirely new class of organizations that possess anti-*kaizen* cultures and are resistant to customer and market choice were to be found, these would be uninspiring to us—and probably immoral. We are open to being wrong and invite such organizations with anti-*kaizen* cultures to stand up and provide us with the opportunity to learn what is good from them and change our mind.

In this book we have defined *kaizen* in its broadest sense as human-centered scientific problem solving and attempt to shed light on its true meaning as a way to dispel common myths about it. We avoid jargon as much as possible, but we continue to use the word *kaizen* because it has become widely accepted internationally and succinct, and we wish to respect and build on the tradition. Although this book is an heir to the tradition of Masaaki Imai's book, *Kaizen: The Key to Japan's Competitive Success*, which brought the important practice of continual improvement to Western business consciousness, we do not claim that Japanese management is superior to management approaches elsewhere. In fact, we show that the *kaizen* approach was brought from the United States to Japan in the first half of the twentieth century and matured there despite the challenges posed by Japanese culture before being reintroduced to the West. In the decades that followed the publication of Imai's book, thousands of organizations worldwide across all sectors studied and applied these practices. Yet, just as Japan's economic miracle was revealed to be fleeting, very few organizations have found themselves thoroughly and sustainably transformed for the better. We believe that in the very near future, by learning from the past with a focus on *kaizen* culture, organizations will achieve these things.

There was far more to Japan's competitive success than *kaizen*, and there is more to *kaizen* itself than activation of improvement teams, the launch of total quality management programs, or the collection of improvement suggestions from employees. These things are necessary but not sufficient. Western management has treated each of these like the proverbial silver bullet to kill the monster of poor performance when, in fact, it is the "culture monster" that must be tamed and turned into the ally of excellence. *Kaizen* was never a silver bullet. A bullet generally can be used only once, whereas *kaizen* must be a continuous process.

Today there is no shortage of books, videos, case studies, workshops, certifications, benchmark visits, and consultants to aid in the learning and application of *kaizen*. The vast majority of professionally managed organizations today are aware of the benefits of *kaizen* or have started various versions of *kaizen* efforts under different names, such as *operational excellence, lean management, six sigma, lean six sigma, total quality management,* or otherwise. Yet very few have sustained the gains they have made, much less built on early success to make human-centered scientific problem solving a sustainable competitive advantage. Total quality management, for example, still vigorously practiced at Toyota and an integral part of the company's management-development system, is largely forgotten in the West or remembered as a program that failed when it devolved into learning and applying a set of quality tools without embedding those practices deeply within the culture of the organization. Although the names of these business excellence programs have changed and their content has been expanded and updated, their mechanistic, results-driven approach to adoption has largely not changed, and the associated failure factors remain.

This brings us to the question of the purpose of this book. We wrote this book because we were dissatisfied with the way the topic of organizational culture is being addressed within the broad domains of process excellence, continuous improvement, and performance management. We must answer the question, "What problem are we solving by writing *Creating a Kaizen Culture?*" Put simply, our problem statement is this: "The success rate for an organization adopting *kaizen* is less than 100 percent." Based on studies and our observations over the past three decades and the authors' combined experience, we estimate that the number of organizations that have attempted and succeeded in creating a lasting culture of improvement is less than 5 percent. On the one hand, this 95 percent failure rate is a major problem. On the other hand, this is a massive opportunity because 19 of 20 firms can still create a *kaizen* culture and reap great rewards.

This book will dramatically improve the reader's chances of success in implementing a *kaizen* culture by closing the biggest gaps in the correct understanding of the following:

1. What *kaizen* culture is and why we need it
2. How everyone, everywhere can practice *kaizen* every day
3. The leader's role in turning *kaizen* culture into competitive advantage

The final missing ingredient must be supplied by the reader: a strong desire to improve based on dissatisfaction with the status quo. It is our hope that the ideas and stories in this book will sew the seeds of healthy dissatisfaction in the reader's mind.

ACKNOWLEDGMENTS

This book would not have been possible without the guidance and professional support of the McGraw-Hill team, including Judy Bass, Amy Stonebraker, Pamela Pelton, Stephen Smith, and Patty Wallenburg.

We would like to thank Jacob Stoller of Conversation Builders for enabling many of the conversations with *kaizen* leaders in this book that help to add color and perspective to the topic of organizational culture.

We thank all of the *kaizen* leaders who contributed generously of their time to share their experience and tell their stories, including Arthur Byrne, Paul Akers, Marta Karlov, Jeff Kaas, Tiziano Toschi, Peter Guerin, Mijo Katavic, Ron Neumann, Emmanuel Dujarric, Mike Matthes, Scott Garberding, Michael Joyce, Mark Graban, Karl Wadensten, Ken Goodson, João Paulo Oliveira, Chris Whittaker, Bob Brody, Keith Jewell, and Joseph Swartz.

We must thank Masaaki Imai for his insistence on the true meaning of *kaizen* as being improvement by everyone, everywhere, every day.

Jaime Villafuerte would like to thank his wife Marilene, partner of fun and work, who makes life more outstanding and rewarding than he ever imagined. To his mentor Walt Garvin, thanks for teaching that the job should never interfere with education.

We thank all members of the Kaizen Institute team worldwide for their dedication to making our founder's vision a reality and providing the experiences and inspirations for many parts of this book. We recognize the particular contributions of colleagues Bruno Fabiano, Steve Burkhalter, Elizabeth Barker, Antonio Costa, Alexandra Caramalho, and Danie Vermeulen.

To everyone who has ever faced difficult challenges on the *kaizen* journey and instead of giving up has asked difficult questions, thank you for inspiration.

MINI GLOSSARY OF TERMS

We have tried to minimize the use of jargon and technical terms in this book. Although a deep understanding of the following terms is not essential to grasp key ideas in this book, this mini glossary will be useful as a quick reference.

3C. A simple frontline problem solving approach in which concerns, causes, and countermeasures are written down in a high visibility area of the workplace.

5S. The workplace organization and problem visualization method characterized by the five actions of sorting, setting in order, sweeping, standardizing, and sustaining.

5 why. The root cause analysis practice of asking "why?" repeatedly (at least five times) and progressively (not five separate questions) until actionable causes are found.

A3. The method to organize problem solving or planning activity on one A3-sized paper, as well as the thinking process which incorporates various *kaizen* principles.

Andon. Japanese for "lantern." A lamp or visual signal used by the people who do the work to alert support staff about problems in the workplace.

Catch ball. The process of back-and-forth two-way communication used during the process of target-setting to come to consensus on achievable plans as part of strategy deployment.

Gemba. Japanese for "actual place." Commonly refers to the workplace where value is added or where customers are served.

Genchi genbutsu. The Japanese phrase meaning "confirm reality with your own eyes" or simply "go see." Literally "actual place," "actual thing."

Hansei. Japanese for "reflection." The act of humbly and honestly looking back at the process and results of our thoughts, feelings, and actions to seek areas for improvement. An important part of the check phase of the PDCA cycle.

Hoshin kanri. Japanese for "policy management." The annual planning process that integrates many *kaizen* principles. Also called *strategy deployment.*

Jishuken. Japanese for "autonomous study" and a common learning-by-doing approach to continuous improvement and management development at Toyota.

Kaizen. Japanese for "improvement." The term in modern business implies continuous improvement, total engagement of the workforce, and valuing small changes as much as larger ones.

Kanban. Japanese for "signboard." In business, any visual signal that limits the amount of work in process (WIP) such as inventories or unfinished projects to rational levels.

Lean. A term describing the condition of organizations which have significantly reduced waste, variation, and overburden across their enterprise by persistently practicing *kaizen* to develop their people and processes.

Muri. Japanese for "overburden," or unreasonable burdens placed on people and processes, which results in waste.

PDCA. The plan-do-check-act cycle of continuous improvement that is the basis of *kaizen.* Developed by Walter Shewhart and Edwards Deming as a practical and continuous application of the scientific method. Also called the Shewhart cycle, the Deming cycle, PDSA (plan-do-study-act), and other names.

Sensei. Japanese for "teacher." A term of respect often used to describe experienced instructors of *kaizen* and lean methods.

Value stream. The sum total of processes from a customer request to fulfillment of that request. A value stream may be defined to extend "from" and "to" various spans such as "order to cash," "concept to launch," or "door to door."

Yokoten. Japanese for "horizontal application." The act of building on good ideas and good practices across an organization, or across different organizations, through deliberate sharing of successes.

CHAPTER 1

Why We Need a *Kaizen* Culture

Change is the only constant.

—HERACLITUS

There are two types of change in this world: change for the worse and change for the better. *Kaizen* is the latter, change for good, otherwise known as continuous improvement. The amount and pace of change in the world today seems to be ever increasing. What we need more of is the will and skill to guide change in the direction of good.

Kaizen enables people and organizations to adapt a set of philosophies and tools towards improving any process, product or service. It is no longer enough to say, "We must improve." We must become better at improving *and* at teaching others to do the same. This requires more than simply raising individual skills in *kaizen*, it requires understanding and overcoming the greatest barrier to continuous improvement: organizational culture.

The degree to which an organization is capable of change will determine not only its performance during good times but also its ability to adapt and survive when external factors erode sales or profit margins or disrupt entire business models. Although there is no known infallible predictor of long-term success, observable day-to-day behaviors that affect decision-making are good indicators of the fate of people and organizations. How we make decisions reveals our character, and character shapes our destiny. This is especially true as the social, technological, economic, environmental, and political factors around us are rapidly changing, requiring that our decision-making processes integrate these new elements ever more quickly. Kotter and Heskett (1992) studied the performance of 207 companies over 11 years and found that those with "adaptive cultures" outperformed the "nonadaptive cultures" with revenue increases of 682 versus 166 percent, net incomes of 756 versus 1 percent, and stock price increases of 901 versus 74 percent. With results such as these, who would not prefer to be adaptive? Yet there are deeply human factors that prevent us from changing how we think and behave. Transforming a culture is far more about emotional growth than about technical maturity. Adaptive cultures support an organization's immediate strategy and short-term business context while also supporting forward-looking plans. Adaptive cultures strive to guide positive change for the long term. We will demonstrate how the core *kaizen* beliefs encourage adaptive cultures through the practice of cooperative, customer-focused problem solving routines.

The characteristics of unhealthy or "nonadaptive cultures" include arrogance, inward focus, and bureaucratic tendencies (Table 1.1). Although such characteristics may support an organization's immediate strategy and short-term business context, they undermine its ability to adapt to change and prepare for the long term. Kotter and Heskett observed that corporate cultures can easily become "nonadaptive," and that changing cultures to be more adaptive was tough but possible with visionary leaders. They predicted that the ability to adapt would become increasingly important in the future. That future has arrived.

What are the cultural traits that promote adaptability? How can we recognize, encourage, and strengthen them? These are questions that humble leaders with a long-term view are asking. The answers require that they lead, shape their organizations, and structure their daily work in certain ways. When these things are done, the result is what we call a *kaizen culture*.

Table 1.1 Adaptive and Nonadaptive Characteristics

Nonadaptive Characteristics	Adaptive Characteristics
Internally-focused, bureaucratic	Customer-focused
Reactive	Proactive
Risk averse	Taking intelligent risks
Information flows with difficulty	Information flows quickly and smoothly
Strong control from the top	Local decision-making and initiative encouraged
Low creativity	High creativity

We will reveal a set of core beliefs and assumptions that underlie the behaviors of excellent organizations. These core beliefs and assumptions generate characteristics such as the Toyota Way and its 14 management principles (Liker 2003). What we call the *kaizen core beliefs* generate the tools, systems, and behaviors that are visible within the *kaizen culture* at Toyota. Too many organizations in the pursuit of operational excellence are in a hurry to copy or borrow the most visible systems and methods that are closely linked to these 14 principles, such as installing red, yellow, and green *andon* lamps to practice the "stop and fix" principle. Without adequately understanding the cultural soil in which these practices are rooted, these efforts do not succeed for long.

In an "nonadaptive culture" within which the avoidance of failure and the need to seek approval are strong and respect for individuals and the encouragement of learning are weak, the trust level may be too low for people to stop work, press the *andon* button, highlight a problem, and call for help. We have seen many well-intentioned workplaces where such lamps were changing colors often, like traffic lights, with support staff moving quickly to the call. We also have seen too many examples where such lamps were unused or blinking red with no response, nothing more than a sad artifact of a culture that does not support exposing and addressing problems by educating and empowering all parties.

As one might expect, becoming an adaptive culture requires that we practice adapting things. Therefore, simply copying best practices from Toyota or another world-class organization will not make us adaptive. We must study with an open mind, gain insight, apply as closely as practical, learn from trial and error, and adapt the method to our own environment. However, these things require that certain values be in place before we even begin, such as curiosity, tolerance for experimentation and failure, and a blame-free environment. These are some of the *kaizen* values that must be explicitly named, practiced, and pursued through daily routines.

Nonadaptive organizational cultures are concerned mainly about their own well-being and not that of others or of the group, are averse to taking risks, and are insular and bureaucratic, all of which result in an inability to adapt quickly to changes in business environments. Adaptive behaviors, on the other hand, demonstrate caring for others, including stakeholders, customers, and employees, and embrace processes that allow them to initiate changes that serve the changing needs of those whom they serve.

How organizations develop and maintain adaptive cultures through the practice of *kaizen* is the domain of this book.

The ABCs of Organizational Culture

We will start with a basic and broad definition of *kaizen* as a people-centered and scientific approach to problem solving for the benefit of society. Next, we must define organizational culture. At the most foundational level, we can say that *culture* is what a group of people or society would recognize as "how we do things around here." In business, we may talk about a culture of teamwork, a culture of corruption, a culture of competition, a culture of nepotism, a culture of secrecy, a culture of openness, and so on. These all describe observable behaviors or how people recognize that they do things. However, culture has many facets and needs to be understood more deeply at a theoretical level before we can adapt it at a practical level to our discussion of a *kaizen* culture.

Henry Ford is credited with saying, "You can think you can, or you can think you can't, and you will be right." Ford was a man who thought he could, and for the most part, he was right. Many people would agree that positive thinking, motivation, determination, and hard work have helped Ford and many others accomplish many great things. What holds true for individuals holds true for groups of people, for organizations. Although what we think and how we think are not magically responsible for creating the reality of our situation, within individuals and within cultures, there is a very real connection between automatic thoughts, beliefs, or assumptions and the behaviors they generate. Beliefs and values create behaviors and actions that result in good or bad organizational performance. A belief that change will never happen and that effort is useless can become self-perpetuating because collective effort is never made and change never happens.

In order to shape our culture and "how we do things" in the desired direction to increase human happiness, business performance, and sustainability, we must look at more than the *how* we do things and move our focus to the deeper levels of *why* and *what*. Our behaviors are *how* we work, our mind-sets are *why*, and *what* is the more tangible and visible elements. These three levels of looking at culture are very useful in our understanding. Professor Edgar Schein's model of organizational culture (Schein 2004) identifies three distinct levels in organizational cultures that he described as:

1. Artifacts, the tangible or visible elements of culture
2. Espoused values, the consciously expressed rules or justifications
3. Basic assumptions, the invisible and unconscious level

The first level of artifacts is most visible, whereas the second and third levels are increasingly less visible to observers. We have adapted Schein's model to reflect our experience that organizational culture within the context of *kaizen* concerns artifacts, behaviors, and core beliefs, what we call "the ABCs of organizational culture" (Fig. 1.1).

At the surface level are artifacts (the "A"), the elements that are most visible or recognizable for those both within and outside the culture. Artifacts of organizational culture could include decor, dress code, or special vocabulary. At the middle level are behaviors (the "B") that are shared and known within the organization and how members of the group espouse or put into action their principles and philosophies. At the deepest level are the core beliefs (the "C") and shared assumptions. These things are often unconscious and taken for granted and are most difficult to recognize or to change.

How is it that organizations develop their own cultures? Schein (2004) observed that when a group of people achieve a certain level of success together over a period of time begin to believe in a set of decisions or actions that appears to have solved problems or led to this success, this creates a set of shared assumptions and builds the foundation of culture. As one would expect, the longer or more repeatedly the people in this organization perceive certain

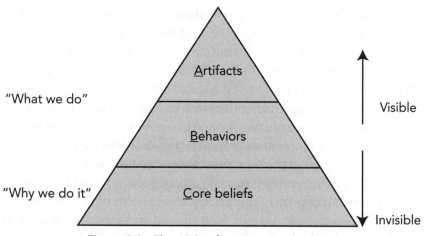

Figure 1.1 The ABCs of organizational culture.

assumptions and behaviors lead to success, the more deeply and strongly these assumptions and behaviors become part of the organizational culture.

If culture is shaped by the repetition of successful practices, why do some organizational cultures hinder high performance? This can be due to a high-performing culture that is well-suited to a certain market situation being exposed to competitive, regulatory, or market conditions that no longer favor the strategies, behaviors, or decision-making processes built into the organizational culture. In essence, the game has changed, but the game plan has not.

The nature of changing business environments argues strongly for the need for organizations to deliberately build adaptive capabilities. If repetition of behaviors that led to success in the past shapes organizational culture, we must question the major assumptions that the behaviors being repeated are indeed those that led to success. A good part of success is the result of "timing meets preparation" or luck. An organization with all the wrong behaviors may succeed for many years because of historical accident, whereas organizations with excellent cultures may be killed off by major economic, social, and historical events, such as war, that are too disruptive to survive even with maximum adaptability and resilience. Truly excellent cultures will not tolerate arrogance and self-satisfaction but will challenge their own success and whether it is by luck, the grace of God, or thinking and behaviors that consistently lead to the making of good decisions.

What Is *Kaizen* Culture?

The notion of *kaizen* that we broadly define as people-centered scientific problem solving directed towards the benefit of society has a very long history with roots in the quality-management work of Dr. Edwards Deming and the programs of Training Within Industry (TWI) from the U.S. Department of War during World War II. We will shed more light on the American roots of *kaizen* in Chapter 2. *Kaizen* culture demands management by fact. This means that decisions must be underpinned by evidence, results of decisions require personal process confirmation by leaders, and time is made to reflect and learn from the results of these decisions, even when the results challenge our beliefs. Beliefs are inherently difficult to change because we hold them to be true in our mind regardless of the evidence and veracity of those beliefs.

Most organizations are not built around incentives for leaders to question their decisions and actions that appear to be responsible for success.

Yet this is what a *kaizen* culture requires. When an organization rewards results without scientific examination of the processes that led to those results, there is great risk for a culture to be built on false assumptions or, at best, assumptions that cannot be proven. This causes organizations to take luck-based decisions or decisions that result in success only when the conditions are similar enough to those when the behavior first succeeded. The reality today is that major changes in external conditions—technological, social, economic, environmental, and political—happen at a seemingly accelerating rate and affect us more directly even when in far-off countries owing to the interconnected nature of our global economies. Therefore, it is essential that we be adaptive in our decision-making, a topic that we will elaborate on in Chapter 4.

We say that an organization has a *kaizen* culture when it values and develops people; builds trust through shared purpose; works toward the long-term interest of all stakeholders; imagines and communicates a positive vision for the future; creates an environment in which the exposure of problems, abnormalities, and inconsistencies is not only allowed but encouraged; treats controlled failures as learning laboratories; follows a common approach to solving problems scientifically; makes decision based on data and facts; holds strong beliefs, assumptions, and values about what is right and good but challenges these to hold up against reality; maintains a sense of humility to seek out and digest foreign ideas and viewpoints; takes intelligent risks; and takes the time to plan thoroughly and build consensus but also acts with a sense of urgency. There are many ways of doing *kaizen* depending on the size and scope of the problem, the number of people involved in problem solving, and speed with which we must test the countermeasures and actions in order for *kaizen* to be effective. We categorize three types of *kaizen* by the cycle of repetition—daily, project, and support—meaning the servant leadership model of management as constant and consistent support for the entire organization (Fig. 1.2).

Although there will be distinct similarities, *kaizen* cultures will not look and feel exactly the same across different organizations. This is one of the greatest misunderstandings of *kaizen*, lean, total quality management (TQM), or any brand of operational excellence: the tendency to treat the surface-level, mechanistic elements of these systems as the entirety. As a result, the typical objection is "we are not Japanese" or "we are not robots" or "we are not making cars," as if a *kaizen* culture anywhere except in

GEMBAKAIZEN®

Daily *kaizen*

The organization

Project *kaizen*

Senior management

Support *kaizen*

Figure 1.2 Servant leadership model of *gemba kaizen*.

caricature is defined by Japanese robots making cars. We humans take in more than 85 percent of the information about our world from our visual senses. When what we are shown as examples of excellence are Japanese factory floors where people, workers and managers alike, to varying degrees perform highly scripted and accurate actions repeatedly, we anchor our impressions of the system on what we see. The key is to look for and recognize what is not visible—the values, beliefs, and assumptions that drive the behaviors, as well as how the visible artifacts reinforce these.

How many leadership teams can look across their organizations and give a real answer to the question, "How much better are we at decision-making today than we were yesterday?" Most would not even know where to go to gather evidence for such a question. Looking at financial statements, the current "gold standard" for measuring performance of for-profit businesses in the current era, will tell us nothing about how much better we are at making decisions today. These measurements of financial performance tend to look back on the results of the past 90 days. For decades, many people have observed that this is like trying to drive a vehicle by looking in the rear-view mirror. This is so because the question is, "How much better are we at decision-making *today*," literally 24 hours after this time yesterday, and comparison with a more remote or abstract past. Only organizations that value visible standards, engage in daily actions to maintain and improve those standards, and link those actions with the annual targets and long-term objectives will have any hope of answering this question. What we just described is the *kaizen* culture in micro.

The *kaizen* process is both top-down and bottom-up. The top-down process begins by imagining an ideal or a target condition and then asking, "What is the goal?" then "What is the current state?" and then "Why is there a gap?" There is almost always a gap between the goal, or desired condition, and the current condition. This is true with the world, within organizations, within our families, and within our own selves. We have met very few people who are totally satisfied in life and seek to improve nothing further. It is human nature to be dissatisfied—to hunger, thirst, and strive. The bottom-up approach is to recognize a problem, a deviation from a standard, and follow the same process of questioning as above. This standard may be far from the ideal, but it is the best-known current condition, method, or way. *Kaizen* guides us toward better ways in the immediate bottom-up fashion and toward the top-down long-term ideals.

Since Masaaki Imai (1986) introduced the concept of *kaizen* into the Western consciousness, we have seen organizations of all types and people from all around the world adapt these principles in order to improve their situation. *Kaizen* itself has and must continually improve, incorporating the latest developments in technology as time-saving tools for both *kaizen* education and improvement-project work. It must incorporate latest discoveries in neuroscience and human psychology, as well as superior models, algorithms, and formulas that arise from the developments of system theory in practice. What we call *kaizen* today matured within one such practical development of a system—namely, the Toyota Production System at the Toyota Motor Company in Japan. Whereas that is a useful benchmark for a *kaizen* culture that has developed and sustained for decades, we must remember that the Toyota Production System that we can observe is an emergent phenomenon of the underlying values and beliefs within the *kaizen* culture. The system is the tree, the financial performance is the fruit, and the culture is everything within the soil below.

We are often asked for industry performance benchmarks. These are mainly useful for shocking an organization into a sense of urgency that it is far behind the competition. This only works when the organization is open-minded, ready to learn, and not prepared with a long list of excuses why "we are different" and the comparison does not apply. An organization that is truly ready to learn from the best has the humility to recognize that it can take good practices from any type organization, regardless of industry or location. What is important is how the people make decisions, learn as an organization, and deliver excellence to their customers. At the level of human behavior, the benchmark is simple: Does everyone across the organization improve every day? Of course, this is a faraway ideal, and there is a long list of supporting behaviors (Table 1.2). We will address these behaviors and the core values that generate them in Chapter 3.

Table 1.2 Traditional versus *Kaizen* Culture Benchmarks

Traditional	Kaizen
Fragmented short-term purpose	Shared long-term purpose
Leaders give direction	Leaders also coach
Lead with power and authority	Lead by example and humility
Leadership enforces rules, bureaucracy	Leadership enables improvements to standards
Go see to catch and punish	Go see to show respect and ask why
Seek out blame	Seek out root causes
Ask for reports	Ask why at the source
Leaders have the answers	Leaders have curiosity
Arrogance, delusion of invincibility	Humility
Respect for profits	Respect for people
Seeking power	Accepting responsibility
Use scale and volume to reduce cost	Reduce scale and volume to expose problems
Hide problems	Expose problems
Work in functional silos	Serve customers in cross-functional teams
Guard internal expertise	Focus on customers
Increase value through addition	Increase value by subtracting the unnecessary
Produce as much as possible while conditions are good	Produce only what is needed now in the amount needed
Use stock to cushion against problems	Use stocks to manage flow
Inspect to catch errors	Inspect to confirm the process

Table 1.2 Traditional versus *Kaizen* Culture Benchmarks (*Continued*)

Traditional	Kaizen
Problem solving by experts and heroes	Scientific problem solving by everyone
Maximize productivity by managing output	Maximize human potential
Unable to criticize openly, satisfied, proud	Dissatisfied, challenge the status quo
Fearing and resisting change	Welcoming new challenges
Management by agendas	Management by fact
Internal competition, backbiting	Winning as a team, healthy competition
Defensive, unsafe environment	Physical, psychological, professional safety
Sporadic improvement	Continuous improvement
Pressure, stress	Tension, laughter
Compliance	Commitment
Optimizing one part at expense of the whole	Optimize the whole—end to end
Group think and conformity	Valuing diversity of perspective
Culture exists to serve itself, separate from business context	Fitting culture to business context, service to customers
Risk averse	Take intelligent risks
Benchmark to critique and justify	Benchmark to humbly learn

But first, let's learn from the interesting case of an American hospital that decided that it needed to copy the Toyota Production System.

Learning How to Adapt from Virginia Mason Medical Center

Many organizations have made the leap successfully from what is visible on the surface to what lies below. Sometimes it is easier for an organization that is entirely different from Toyota to do this. The Virginia Mason Medical Center in Seattle, Washington, is one notable example. In 2001, the hospital set out on an ambitious system-wide program to improve patient safety, quality, and cost. The hospital's approach was to employ both Japanese and

American lean manufacturing consultants and to take dozens of its physicians and administrators to Japan to observe what Toyota does in its factories. The participants returned and created the Virginia Mason Production System (Fig. 1.3). This is notable for the fact that the senior leadership at Virginia Mason deliberately called it a "production system" in order to break the mind-set that "we are different" and "*kaizen* does not apply to hospitals" and that the system of patient care was a set of processes no different from production to which *kaizen* principles could be equally applied. Considering the sense of urgency facing Virginia Mason with its struggle for financial survival, the aggressive and even shock-inducing approach of CEO Gary Kaplan is understandable.

It should be noted that the Virginia Mason Production System was not a copy of the Toyota Production System. It was not a formulaic application of improvement tools and systems such as 5S, *kanban* (demand-based replenishment signals), and flow, although these were used whenever they made sense. Rather, as the pyramid diagram in Figure 1.3 shows, it was a

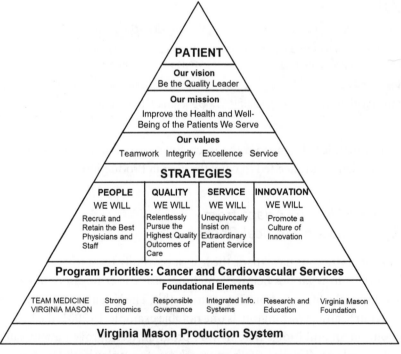

Figure 1.3 The Virginia Mason Production System.

system based on principles and behaviors, with the tools and artifacts acting as enablers. The professionals working at Virginia Mason were encouraged to gain a deep appreciation of these principles and apply them to the processes across the medical center. As a result, the following shifts in culture occurred (Kenney 2010):

▲ From expert-driven medicine to team medicine
▲ From physician-centered to patient-centered medicine
▲ From fear of raising safety alerts to comfort
▲ From an acceptance of the complexity of clinical pathways to an insistence on smooth flow
▲ From "we've always done it this way" to "there is a better way"

The Virginia Mason Production System is now a global benchmark for cultural and performance transformation in healthcare. The Virginia Mason Institute offers training, hands-on learning, and certification to others who wish to apply the lessons from the Virginia Mason Production System to their own hospitals. What is the secret of the sustainability of the Virginia Mason Production System? One of the keys, according to Rodak (2011), is the culture of collaboration and respect for people created by the leadership. Steve Schaefer, vice president of finance for Virginia Mason, states: "From the CEO to the CFO and COO to vice presidents and directors, there is mutual accountability that requires us all to be engaged in this work in all the various areas that we lead." Another key factor was the constancy of purpose and long-term focus demonstrated by leadership. "We kept waiting for leadership to flinch, because at first we thought it was just another management thing. But they didn't waver," says Schaefer. In addition to support and direction from top management, bottom-up engagement was critical. "Culturally, it started from the top down, but our objective is to see this really as a grass roots effort at wanting change coming from the frontline staff," says Schaeffer.

A hospital in Seattle, Washington, could hardly be more different culturally than an automotive factory in Toyota City, Japan. Yet we have a success story that is well-documented, and the participants openly share what they have learned with the world, expanding *kaizen* beyond their organization and into the broader community. Why, on the other hand, are there so few such shining examples decades after the Toyota Production System know-how became public and consultants became available as teachers? Is Virginia Mason a statistical freak, or are there factors holding back the other organizations that embark on similar journeys? We believe that Virginia Mason, like other organizations we profile, was uniquely successful not by chance but because of the fact that it aligned the change effort with its long-term purpose of patient centricity and service to the community, converted urgency into positive motivation, built on a foundation of intellectual curiosity and valuing the scientific method that was a basic belief of the staff, and did not flinch in the face of tough decisions.

Kaizen Culture as Countermeasure to Extension Transference

Another important concept that may help to explain why organizations struggle to establish and sustain performance excellence comes from the work of American anthropologist Edward T. Hall. There are two parts to the notion of *extension transference* as introduced by Hall (1976). Extensions are things that we humans use to overcome challenges and improve our lot, including language, tools, technologies, social structures, and institutions. These are either created by human minds and hands, such as tools, machines, and institutions, or emerged with the development of our species, such as language. In the case of language, sounds represent ideas, but are not the ideas themselves, they are extensions. We also have written language, which is yet another extension of spoken language.

Extension transference is the idea that over time the extensions that we create to help us to fulfill a purpose become fragmented and disconnected from their original purpose. In the worst case, we lose sight of the original purpose of the extension such as a tool or institution. We can easily recognize many examples of extension transference within modern organizations supposedly dedicated to continuous improvement and business excellence. There are many KPIs (key performance indicators) that are intended to measure results but that drive the wrong behaviors, such as quarterly earnings reports. Another example is a problem solving document template that is intended to guide the thinking process, but instead becomes a mindless fill-in-the blank exercise. In extreme cases, *kaizen* activity itself becomes the end and not the means to an end, with the number of *kaizen* events, *kaizen* suggestions or *kaizen* events becoming the goal rather than profit and people development as the goal.

Hall could easily be speaking about modern federal government, modern functional organization structures, or dysfunctional continuous-improvement programs when he writes that it is "the complexity of our systems that distracts us and prevent us from remembering the original purpose of our systems—their most basic goal." Extension transference causes us to forget the point of the complex systems we set up to serve us. We build a machine to serve us, but find ourselves working for the machine. When realized as living the values rather than using tools, *kaizen* maintains alignment with long-term purpose and continually engages people through top-down target setting, bottom-up idea generation, and problem solving dialogues on the front lines. When *kaizen* is employed superficially and the improved financial performance to benefit a few becomes an end in itself rather than a means to fulfill the higher purpose of developing people, creating prosperity for many, people become alienated, and *kaizen* is no longer change for good. When we forget that while the tools guide the improvement, it is people who make continuous improvement, suboptimization of continuous improvement occurs, and we have extension transference.

When we see organizations pursue continuous improvement only with the aim of cost reduction or only to measure the number of *kaizen* events held or the number of black belts certified, these are examples of misalignment from purpose. Unless we practice reflection and periodically check that our tools, systems, and programs are still serving their purpose, we are at risk of becoming alienated even from the most brilliant inventions and methods of humanity. We begin to feel powerless and alienated because our actions are not aligned with purpose. Organization-wide change programs, whether they are driven by short-term financial performance needs, long-term human-development objectives, or both, are at risk of extension transference, especially when these changes move more quickly and begin to feel like they are taking on a life of their own. *Kaizen* is not change for the sake of change or even just change for the sake of improving business results—it is also change in order to develop people. In order for it to be sustainable, the motivation to develop people must come not only from a true caring and respect for individuals and society but also because better people and better thinking bring better results. Studies and experience increasingly show this to be the case. *Kaizen* challenges us to constantly ask, "What is the purpose of this process?" and seek better ways to fulfill this purpose. This behavior of seeking the answer to this question counteracts the risk of extension transference, and is critical to creating a *kaizen* culture.

Human Adaptability, the Strategic Competitive Advantage

Whether we think of it this way or not, we all are on a team, one way or another. Imagine that you are on a team, competing against other teams. What if there were one team that seemed to learn and improve more rapidly and consistently than the others? This team is not the most popular or glamorous team, but it wins consistently, perhaps by small margins. Not only that, this team does not hide the secrets of its success with competing teams. What if your and other teams had attempted to copy this or that part of the winning team's strategy, but finding that this was not enough, concluded, "We are different. Their strategies will not apply in our team's culture"? What if year after year this continues, the winning team keeps winning, and the competing teams ignore the obvious and try this or that new strategy, making progress some years, falling back in others? This has been the true history of the automotive industry since the decades surrounding the turn of the century. Liker and Morgan (2006) write of Toyota's learning processes embedded within its management system: "Toyota's awesome ability to learn quickly and improve at a regular cadence may well be the characteristic of Toyota its competitors should fear most." Luckily most of us do not have to compete with Toyota. However, all of us, including Toyota, face an even more terrifying competitor: the monster of our own culture.

Several studies substantially underpin our experience about the need to create a *kaizen* culture to achieve sustainable superior performance. Pfeffer (1994) makes a solid case that the source of competitive advantage of successful companies such as Southwest Airlines is their culture. Southwest was able to be the top stock performer, with a 19,907 percent return, from 1972 to 1992 in an industry "characterized by massive competition and horrendous losses, widespread bankruptcy, virtually no barriers to entry, little unique or proprietary technology and many substitute products or services."

Southwest's outstanding performance has lasted. In 2012, the company was the only U.S. airline with 40 consecutive years of profitability. The same year, it was ranked again number one, with the lowest number of customer complaints of all U.S. airlines by the Department of Transportation (DOT), a position that it has had since 1987, when the DOT started publishing airline customer satisfaction data. Pfeffer's well-documented analysis concludes that Southwest's competitive advantage in productivity and exceptional service level "comes from its very productive, very motivated, and by the way, unionized workforce." Pfeffer recognizes that "the culture and practices that enable Southwest to achieve its success" are sustainable and difficult to imitate by competitors. Pfeffer (1998) also provides more examples of successful companies achieving superior performance by putting people first rather than looking outward for the winning strategy. He provides relevant data to make the case that "senior managers of the most successful firms worry more about their people and about building learning, skill, and competence in their organizations than they do about having the right strategy" and warns that "the conventional wisdom about sources of sustained success is wrong. Companies do not have to be large, do not have to go through waves of downsizings, and do not have to be technologically sophisticated, the market share leader, or even global to enjoy substantial economic returns." Pfeffer provides seven dimensions that constitute key people-centered management practices for sustainable economic performance.

Barlett and Ghoshal (2002) argue that companies need to "accept the challenge of creating an environment that will attract and energize people" to create a sustainable competitive advantage. "Today managers must compete not just for product markets or technical expertise, but for the hearts and minds of talented and capable people." While these authors don't explicitly use the term *kaizen core beliefs*, they are describing them and their effect when they are applied.

Despite strong academic evidence for the benefits of developing organizational culture as a competitive advantage, many companies continue to focus on strategies based on positioning and acquisition. Perhaps the world's best-known business-school-professor-turned-consultant, Michael Porter wrote in *What Is Strategy?* (1996) that "continuous improvement . . . tools unwittingly draw companies toward imitation and homogeneity." He made the case that "total quality management and continuous improvement" can't provide a sustainable competitive advantage. Porter's thinking is sadly flawed owing to a mechanistic view about continuous improvement that is common in Western organizations. Porter wrongly assumes that continuous improvement is about *making things better* instead of *making better people*, confusing cause and effect. Toyota's statement, "We don't build cars, we build people," is often cited as a uniquely Japanese viewpoint toward long-term commitment to employees. Nearly a century ago, Reverend Samuel Marquis, who headed Ford's employee relations department, wrote about the misconception that Henry Ford was in the automobile business (Grandin 2010), saying: "It isn't true. Mr. Ford shoots about fifteen hundred cars out of the back door of his factory every day just to get rid of them. They are the by-products of his real business, which is the making of men."

As a pillar of *kaizen* and the Toyota Production System, the TQM tradition also reflects the true meaning of continuous improvement, which is to develop people. As a result of making better people into better decision makers and better problem solvers, better processes and products are

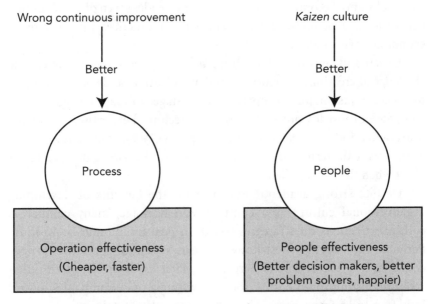

Figure 1.4 *Kaizen* culture focus on people.

created (Fig. 1.4). This is the meaning of *total quality* in TQM. However, trying to put in place a culture and organization that makes better people is a harder and longer-term proposition, not an easy sell for a strategy consultant. Porter's shallow dismissal of continuous improvement and promotion of easy-to-adopt positioning strategies may appeal to leaders oriented toward short-term profit and may explain the popularity of his ideas. By contrast, each year, more committed companies have chosen to go down the harder path of better outcomes through better people, and in this book, we will introduce many encouraging examples—as shown below.

Barlett and Ghoshal (2002) argue that organizations must place human resources and the development of human capital at the center of their strategy. In their view, the role of management becomes the creation of policies and processes for recruiting talent that best fits the organization's strategic needs. Moreover, management must coach and develop these new recruits continuously, treating them as long-term investments. Human resources professionals have typically responded to the increasing importance of human capital with new approaches to recruitment, innovative training programs, and experimentation with compensation packages. However, these are tools available to everyone, and these tactics are seldom taken as point solutions rather than a total system transformation to develop adaptive, learning organizations. Even the best-selected, best-trained, and best-compensated talent will not achieve its full potential within a culture that does not have values aligned with the human-capital strategy outcomes it desires. We demonstrate that the *kaizen* core beliefs of continuous improvement and the respect for human potential go hand in hand in developing capable people through the process of building capable business processes.

Futurist Alvin Toffler (1991) wrote that knowledge, not military strength or financial resource, would become the true source of power in the near future. The world has seen many of Toffer's predictions play out. Particularly relevant to our thesis of the need for a *kaizen* culture is Toffler's insight that "[t]he illiterate of the 21st Century are not those who cannot read and write but those who cannot learn, unlearn and relearn." More than being illiterate, organizations and leadership teams that are unable or unwilling to adapt to changes in the world by letting go of habits, business models, or structures that are not suited for the environment, and unable to adopt new ones are at risk of being replaced by more agile and adaptive competitors.

Is *Kaizen* Culture Easier Within Japanese Culture?

Artifacts are the products that emerge from human actions as a result of their cultural assumptions, values, and beliefs. Although artifacts have a reinforcing effect on assumptions and behaviors, they seldom immediately cause them. A culture develops or changes because of shifts in values or attitudes. These are influenced by social, technologic, economic, or historical factors. We can say that jazz music is an artifact of twentieth-century American culture. This artifact was a result of people wanting to make music that sounded good to them and satisfied other basic and universal human needs, such as food, shelter, and self-actualization. Jazz was very much a product of the culture of the times.

In the story of why jazz happened, Marc Myers revealed that a combination of social, technologic, economic, and political events, including slavery, the emergence of recorded music, World War II, the rise of radio, the civil rights movement, the dominance of record companies in the music industry, and so forth, shaped peoples' values and constrained or inspired their creative energies, resulting in the development of jazz. In turn, as jazz became part of the culture, it influenced the assumptions, values, and behaviors of many people who came into contact with this artifact of American culture during the middle part of the twentieth century. Today, popular culture in the form of entertainment and the arts has worldwide influences on culture and even informs and shapes beliefs and behaviors. These cultural artifacts can be very powerful. In a similar way, the artifacts of *kaizen* cultures—the tools, routines, visuals, and practices—influence our thinking and therefore our organizational cultures.

Jazz was shaped by many historical factors. Although jazz has influenced and was influenced by American culture, there is nothing about living within American culture that makes jazz easier to appreciate, any more than jazz requires that we be of African origin to connect with its rhythms or of European origin to play its instruments. People all over the world appreciate listening to and playing jazz. It would be nonsense for musicians in one country to say, "Jazz will not be appreciated here. We are not Americans." We will show that much as jazz was a product of American history, *kaizen* was a product of historical and economic factors rather than Japanese culture. *Kaizen* is an artifact and emergent phenomenon resulting from a set of what we call *kaizen core beliefs*, many which are common to all humanity and all of which can be adopted by anyone. We will show how *kaizen* activity influences the deeper values and assumptions of its practitioner, creates belief in change, builds trust and relationships, encourages respect for people, strengthens the will and skill for solving problems collaboratively, and creates community—addressing basic and advanced human needs.

To the heartbreak and disappointment of many who arrive to lead organizations in Japan, *kaizen* culture is the exception rather than the norm in that country. Frequently we have heard the objection that "*kaizen* will not work in our culture because we are not Japanese." In fact, it is amazing that as we help more multinational companies adopt the *kaizen* approach to improve their business processes and develop their people, it becomes clear that one of the most underappreciated costs of operating a multinational business is the cost of complexity due to culture differences. *Kaizen* can create a common language for managing and improving processes, developing people, and exposing problems in a way that is independent of the context of local cultures. But this requires understanding the unique differences of local cultures, such as the degree of group orientation versus individualism, authority-based decision-making versus consensus decision-making, avoidance of failure versus tolerance of failure, explicitness in communication versus ambiguity, and so forth. It was Japan's history, not its culture, that enabled the emergence of *kaizen*, a result of Japan's response to the need to adapt. Not all Japanese companies adopted *kaizen*, adapted, or survived. There were plenty of cultural barriers in Japan to *kaizen* culture, some simply part of human nature. We will see in Chapter 2 how *kaizen* developed and where the culture of Japan helps or hinders becoming more adaptive.

Takehiko Harada joined the Toyota Motor Company in 1968 and served for over 40 years, including five years as president of a supplier company. He learned the Toyota Production System hands-on while serving in various management and director roles in Toyota factories. Like many of its Japanese leaders, he faced challenges in bringing Toyota's management approach to cultures in other countries. Harada uses the analogy of a three-story building to compare the cultural differences one must be aware of to successfully manage overseas operations (Harada 2013). There are actually four levels, including the "ground" or foundation on which the building is constructed (Fig. 1.5).

The experience of Harada and others in attempting to implement the Toyota Production System outside Japan shows that trying to make changes

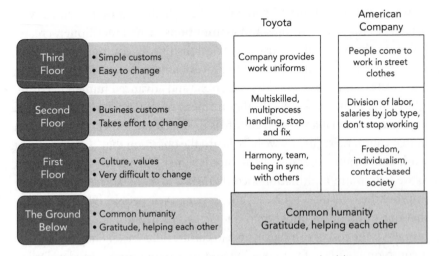

Figure 1.5 Company culture as a three-story building.
(Adapted from Harada 2013.)

to the "first floor" values and beliefs as a requirement to implement the Toyota Production System in the United States does not work so well. Making changes to customs and practices at the second-floor level requires understanding and respect for the local values and culture on the first-floor level. For example, following up and checking on progress of a subordinate to an assignment are experienced by Americans as an invasion of privacy and an act of distrust in the subordinate's competence. In the Toyota culture, checking on one's subordinates is an act of caring, coming from a belief in the responsibility to develop the individual. These two experiences are based on very different cultural assumptions. Given this understanding, Harada advises Japanese managers to stop this behavior and never "look over one's shoulder," using the English phrase he learned. There are many such cultural differences that result in misunderstandings and that act as a drag on performance, and this is true company to company, not only between countries.

Anthropologist Edward T. Hall introduced useful tools in understanding cultural differences in many areas. According to Hall (1976), *high-context culture* and the contrasting *low-context culture* refer to a culture's tendency to use less or more information in routine communication. This choice of communication styles translates into a culture that will cater to in-groups, an in-group being a group that has similar experiences and expectations, from which inferences are drawn. In a high-context culture, many things are left unsaid, allowing the culture to explain. Words and word choice become very important in higher-context communication because a few words can communicate a complex message very effectively to an in-group (but less effectively outside that group), whereas in a lower-context culture, the communicator needs to be much more explicit, and the value of a single word is less important.

These are relative and not absolute qualities, and the reader may recognize that even within regions of his or her country there are differences, such as the southern United States, which is higher context than the North. Even within a low-context culture there will be high-context groups with their own "in jokes" and shared understanding that is closed to outsiders of that group. As with other cultural assumptions, the key is to make the invisible beliefs visible so that they may be understood and behaviors can be shaped in the desired direction through the practice of routines, problem solving, and many small victories.

Rather than abandoning practices such as "looking over someone's shoulder" entirely, it is better to expose the cultural assumptions, understand the thinking process, and find common ground with the American who feels that his or her ability to do assigned work is being questioned or his or her privacy is being violated. Why not explain the genuine desire of the Japanese manager to help the worker grow? Because Japan is a high-context culture, Japanese managers do not expect to need to explain everything in detail. America is a low-context culture, preferring the opposite. As a result, one side wonders why the other side doesn't "get it," and the other side wonders why there is no explanation for a behavior perceived as rude. The interaction of the three levels of culture here are

1. *Visible behavior.* Following up on assignments.
2. *Values behind the rule.* Superiors must mentor and develop subordinates by checking their process and results of their work.
3. *Basic assumption.* Both teacher and student understand and accept these roles within the organization.

Even if points 1 and 2 are explained and linked, without a belief in point 3, the American may continue to resent being coached when he or she feels that there is no need. The nervousness at being silently observed is a general condition of humanity at the ground-floor level, on top of which Toyota people have built the practice of observing and being observed across decades. There is no Japanese cultural assumption that makes Japanese people comfortable with having superiors "look over their shoulder" or monitor their work.

Another example Harada gives is the reluctance of American workers to stop work and call for help to address a problem found because of a fear of punishment. This would seem to be a business custom (second floor) rather than a national culture (first floor), yet this was one of the most difficult things to change. We can speculate that this is so because fear is at the deepest level of "common humanity," even though the custom itself is superficial and learned in a very specific environment, the workplace. Harada writes for a Japanese audience, and his conclusions are biased toward the Japanese perspective in pointing out what does not work in the United States, without giving examples of how local cultural values actually make implementation of the Toyota Production System easier. Things look different from an American or bicultural perspective.

Nonetheless, Harada's premise and advice to be aware of the level at which the proposed changes will affect people's customs are quite appropriate. When the new way of working is built on a solid foundation of common humanity, for example, people who care for each other and people who want to find value and purpose in their work and are grateful for their opportunities, the persistent manager will succeed in building a culture that excels. Before the reader can fully answer the question posed here—of whether adopting a *kaizen* culture is harder within his or her country and company situation—we must explore more deeply the true meaning of *kaizen* and its surprising American origins.

CHAPTER 2

The True Meaning
of *Kaizen*

A business that makes nothing but money is a poor business.

—HENRY FORD

As a pioneer of modern mass production, Henry Ford observed and commented on the direct relationship between the amount of time his products stayed in the process of manufacture and the cost of those products. His flow production method shortened the production cycle, increased outputs, and kept prices low. Ford also was the model for the *kaizen* idea suggestion system at Toyota. It would be a mistake to equate the genesis of *kaizen* at either Ford or Toyota simply with a desire to shorten production cycles and reduce costs. The Ford Motor Company certainly made Henry Ford a lot of money, but that was by no means the only thing he wanted to make. He wanted to make a difference in the lives of his employees and have what he saw as a positive impact on society.

Henry Ford was an industrialist but also a social reformer who wanted to see his employees develop their character based on his ideals. The Ford Motor Company faced the problem of high turnover of its workforce; as people struggled or failed to adapt to the assembly-line work culture, Ford addressed the problem through a combination of higher pay and education (Meyer 1980). By more than doubling minimum wages from $2.35 to $5 per day, Ford attracted the best people and reduced turnover. However, Ford also demanded that the workforce participate in various activities to develop their habits and character according to his vision. This included English-language education for immigrants to the United States, as well as investigations by the Ford Sociological Department into the "thrift, honesty, sobriety, better housing and better living generally" of each worker. In a world today that recognizes the value of cultural and linguistic diversity within an organization as a competitive advantage, Ford's actions may appear paternalistic. Some of the more intrusive and heavy-handed practices of the Ford Sociological Department were later relaxed or stopped. Put in the context of Ford's day, he was also a pioneer in recognizing the importance of shaping the values and character of the people in his organization in order for the Ford Motor Company to operate its business model successfully.

Such good intentions did not always meet with success, nor were they always welcomed by the people whose culture Ford wanted to change. Henry Ford's misadventure in the Brazilian Amazon began as a plan to grow rubber to supply his factories in order to break the rubber monopoly. Called *Fordlandia*, his vision developed into Cape Cod homes, hamburgers, dress, cultural events, and prohibition imposed on the native population (Grandin 2010). The workers rebelled against living and working conditions incompatible with local conditions, and the entire venture failed spectacularly, without supplying any rubber to the company. Henry Ford never visited Fordlandia.

There are many lessons that we can take from the failure of this wealthy, powerful, intelligent leader to bring about culture change. There is the importance of humility, of understanding of the process deeply enough to adapt to local conditions, respect for people as individuals, and the hazards of taking grand actions on the foundation of weak consensus. Ford met with success during his career in making others adapt to his vision and left us with many artifacts, but the culture he created was not adaptive.

At the middle of the twentieth century, Toyota was by all measures the underdog. It did not have volume to enjoy economies of scale. The company lacked the financial resources to pay the highest wages, and it did not have a large industrial base from which to draw human capital. The company did not have resources for grand experiments. The need to change at Toyota was born from urgency and scarcity. What the leaders of Toyota did have in common with Ford was a desire not only to successfully build cars and make money but also to build people and make a better society. From this deep belief in the importance of developing the skills and character of employees at Toyota, the unique moral characteristic of *kaizen* was born.

The True Meaning of *Zen* in *Kaizen*

Some people who have not yet appreciated the full set of core beliefs underlying *kaizen* culture may feel that *kaizen* is an oppressive, never-ending process of continuous improvement focused on finding the ever-smaller increments of improvement. Perhaps because of its introduction to the West being primarily through manufacturing and industrial engineering channels, a false belief in the incompatibility between streamlining processes and individual creativity has built up. Although the *kaizen* process is scientific, it is also people-centered in its values and approach. There are built-in beliefs and biases in *kaizen* that guide change toward what is good, right, and moral. Engaging the full potential of people, including their creativity, toward fulfilling a higher purpose is good. *Kaizen* is often defined as "change for the better" or "improvement," but in order to understand the assumptions built into a *kaizen* culture, we must look a bit deeper into the meaning of the word *good*.

Kaizen is a Japanese word that is comprised of two *kanji* (Chinese ideograms). These ideograms developed from sketches mimicking things found in nature over thousands of years. Although the components of the ideograms are often recognizable to those in Asian culture who use them, it is not often that people pause to consider the origins of these words or what the individual components mean. Ideograms for simple ideas or things resemble the objects they represent, such as those for *mountain* or *tree* (Fig. 2.1).

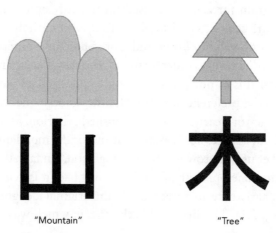

"Mountain"   "Tree"

Figure 2.1   Ideograms for mountain and tree.

卜 = whip, club, branch

又 = right hand

...combining 卜 and 又

攴 = to strike, to whip

㠯 = snake-like creature, evil spirit

...combining 㠯 and 攴

攺 = an early form of 改

改善 = kaizen

Figure 2.2 *Kai* = "driving out the bad."

There is no ambiguity that the ideogram for *kai* means "to change, to remove the old with new" from a very early Chinese ideogram for the action to strike away pestilence, curses, snakes, or evil spirits, according to Tang Dynasty government official and scholar Li Yangbing (Fig. 2.2).

The ideogram for zen has caused more confusion. First, it is often mistaken for the *zen* of Zen Buddhism. That is an entirely different word coming from the Sanskrit *dhyana* and represented by a different ideogram and means "meditation or contemplation." The zen of *kaizen* means "good." However, even this is not precise enough. The English word good can describe subjective preferences, such as "It tastes good to me" when referring to food; accuracy or nearness to an established criterion, as in "That was a good shot" about a successful basketball throw; or the moral quality of being right, as in "Let us not grow weary of doing good." In fact, the meaning of zen is the latter. The opposite of zen is not just bad, it is evil. According to Shirakawa (2006), there are three parts to the original ideogram for the zen in *kaizen* (Fig. 2.3). The first part is the sheep, which was the sacrificial

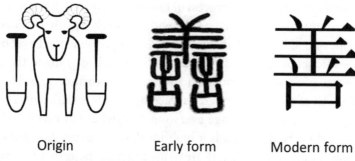

Origin   Early form   Modern form

Figure 2.3 The true meaning of *zen* in *kaizen*.

animal to the gods. The second part is the square container that represents a bowl or urn used to offer prayers to the gods. The third is a needle that was used for tattooing. Combined, the ideogram tells the story of "trial by sheep" from ancient China. Both the accuser and the accused would place their vows in the containers, agreeing to the punishment of tattooing should they be found guilty. In the process of being sacrificed, the sheep would render judgment on who was right between the accuser and the accused.

A *kaizen* culture not only must be adaptive, but it also must have a built-in moral compass, a tendency to take the direction that increases greater good in the long term. What we will introduce as *kaizen* core beliefs in Chapter 3 underpins this bias toward virtue, with core beliefs such as respect for people, serving others, and an alignment with a long-term purpose. *Kaizen* culture is inherently moral with its people-centered focus on directing energy and effort away from areas that do not serve the higher purpose. At the most basic level, this is expressed as wasting less time, energy, and resources and increasing value to the customer. If we forget the true meaning of *zen* in *kaizen*, though, even the successful activity to remove waste risks becoming another example of Hall's extension transference, with good meaning only "efficient" but perhaps not meaningful or good for people.

A Reintroduction to *Kaizen*

Most readers will by now have a basic awareness of *kaizen*, and many will be advanced practitioners. To some, *kaizen* is synonymous with the five-day *kaizen* event or *kaizen* blitz, a process of bringing rapid and dynamic change with a cross-functional team, typically within a week. To others, *kaizen* is a philosophy of making small, incremental daily improvements. To yet others, it is a specific set of daily behaviors and actions such as going to the actual workplace to observe, to clean up and organize the workplace, and to make work flow. All of these are correct, but each by itself is insufficient. The true meaning of *kaizen* is to engage everyone everywhere in making change toward good every day. This is a high ideal, one worthy as a pursuit of continuous improvement. There are many practical steps that we can take in this direction, and "everyone, everywhere, every day" must not be excused as unattainable perfection, lest complacency set in with "most people, most places, most days" doing *kaizen*. While 51 percent may be a majority, it is far from a critical mass.

The development of *kaizen* as an integral part of the management system of certain Japanese companies began not with an intent to create a particular business culture, but rather from a business need. As such, it was very organic and not programmatic. The adoption of *kaizen* among companies around the world has been through a combination of need-driven organic processes of self-study and trial and error and highly structured corporate programs to rapidly copy best practices or benchmark systems of industry leaders. Both approaches have their strengths and weaknesses, and neither should be dismissed out of hand. Although the starting points and rates of progress vary widely, the structure of *kaizen* transformation initiatives adopted by successful organizations tends to arrive at a similar place. We identify three major types of *kaizen* by cycle of activity and their purpose (Table 2.1).

Daily *kaizen* includes all short-cycle improvement activities, including *kaizen* suggestions, natural team-based problem solving on the front lines,

Table 2.1 The Purpose of the Three *Kaizen* Cycles

Purpose	Daily *Kaizen*	Project *Kaizen*	Support *Kaizen*
Strategy and change	Third	Third	Main
Results and change	Second	Main	Second
Process and learning	Main	Second	Third

and maintenance of standards—to be detailed in Chapter 6. The main purpose of *daily kaizen* is learning and reinforcing values, and daily *kaizen* is absolutely critical to long-term sustainability. The direct financial benefits are relatively small compared with the other two major types of *kaizen*, and it is rare that daily *kaizen* has a large impact on the direction or strategy of an organization. *Project kaizen* encompasses the activities of temporary teams working on *kaizen* events, six sigma projects, business-process redesign, new-product introduction, and other project-based improvements. The main purpose of projects is to achieve rapid performance improvement and financial results, with strategic and learning objectives typically following next.

Support kaizen includes all strategy development, planning, talent recruitment and development, training and certification, motivation, and recognition, and such management activities are required to steer and guide the continuation and success of *kaizen*. By *support*, we mean senior management, based on the servant-leadership model of *gemba kaizen* introduced in Chapter 1. The main purpose is strategy and direction, with a secondary focus on making sure that results are being achieved through daily *kaizen* and process *kaizen*, with an emphasis also on reflection and learning. It is through the judicious combination of these three cycles of activity that *kaizen* becomes a natural part of how the work gets done, and "everyone, everywhere, every day" becomes possible. There are no bright lines separating the three types of *kaizen*, and although all three contain necessary activities and elements, they must be adapted to suit each company's situation.

Why did we develop so many different understandings of *kaizen* decades after first introducing this practice to the West? Perhaps this is so because we label things the best we can based on what we see and understand. When *kaizen* was first practiced, the many small improvements were the easiest to see because they were many, concrete, and often visible in the workplace. Larger improvement projects are by nature fewer and harder to observe in day-to-day operations. An observer on a Japanese industrial study tour in the 1980s would note how clean, organized, compact, and free of defects the factories were and be shown the many before and after photos of *kaizens* made by employees. Frequently these were quite small improvements, reducing time by mere seconds or motions by a few centimeters. We need to remember that the Japanese companies being benchmarked had been practicing *kaizen* for decades by the 1980s. Most of the so-called low-hanging fruit had been picked, and the remaining items were either high-impact improvements that took longer to implement because of either cost or complexity of the topic or small *kaizens*. Longer-cycle problem solving that happened through dialogue, coaching, checks, and reviews over a period of weeks could only be observed by being a part of this process. The accumulated impact of developing people could not be readily observed during a short period of study.

It is no surprise, then, that the *kaizen* ideas observed were small, many, and incremental. To others who were introduced to *kaizen* after the 1990s by Japanese consultants in the business of leading *kaizen* events, the week-long format became the definition. None of these by themselves are the correct definition or totality of *kaizen*. Today, the understanding of *kaizen* has expanded to include the small *kaizen* suggestions, day- or week-long cross-functional *kaizen* events, practical problem solving on the front lines supported by management as coaches, often called *A3 thinking* or *A3 problem solving*. But these are only forms of *kaizen*, different cycles of activity based on the scope of the opportunity. What is important is to remember the common thread—that all types of *kaizen* serve to deliver results and to develop people. Although these are timeless themes, there is a specific time and place where the seeds for what we know as *kaizen* today were planted and grew—the dramatic meeting of two cultures, Japan and America, in the mid-twentieth century.

American Roots of *Kaizen*

After World War II, with the aim of helping to rebuild Japanese industry and the national infrastructure, American occupation forces brought in various American experts. The Japanese were eager to learn the various methods to improve quality and productivity while reducing inventories and capital equipment–related costs. The American occupation needed a stable high-quality manufacturing base in Japan to produce radios and other basic necessities for communication and nation-building by an occupying force. The historical environment of scarcity led to the need to

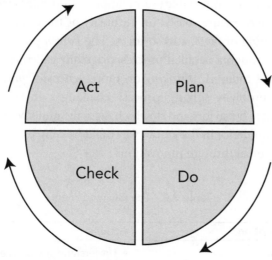

Figure 2.4 PDCA cycle.

waste as little as possible and to find creative solutions to problems. Among others, statistician and quality expert W. Edwards Deming was invited to Japan to train local managers. This led to local efforts to study and implement quality-control (QC) tools and the plan-do-check-act (PDCA) cycle of problem solving, often called the *Deming cycle* (Fig. 2.4).

In particular, the phases of the PDCA cycle deserve careful study and appreciation. At the simplest level, it may be sufficient to go through all four phases of plan, do, check, and act/adjust when solving problems or implementing business plans. However, without being prescriptive, it is important to demand a level of rigor in the methods that are followed during each phase. For example, what does it mean to plan? Charles Kettering, head of research for General Motors (GM), is credited with saying, "A problem well stated is half solved," and in fact, a more powerful statement would be, "A problem poorly stated is virtually unsolvable."

The deep examination of problems, through investigation by direct observation, the building of consensus on the nature of the problem, and analyzing root causes through scientific investigation are not part of human nature. Humans like to jump to solutions. Other typical behaviors include avoiding problems or sweeping them under the rug to avoid blame, rushing to a "fix" to avoid uncomfortable examination of the underlying root causes, selecting countermeasures based on authority or the loudest voice rather than the weighing of facts, and so forth. The *kaizen* process deliberately takes people through a detailed process, originally called the *QC storyline* (Table 2.2) and today *A3 thinking* or, more generally, *practical problem solving*. The problem solving process challenges these unproductive solution-oriented behaviors and slowly changes the underlying assumptions by building confidence in the *kaizen* way of addressing problems as a team and achieving breakthrough innovations.

Table 2.2 The QC Storyline

Step	Description	Actions
1	Theme selection	• Understand the problem points • Select the theme
2	Grasping the current situation and setting targets	• Grasping the situation by collecting facts • Determining the key characteristics • Set targets (what by when)
3	Creation of action plan	• List the actions • Assign responsibilities and dates
4	Analyze root causes	• Investigate the key characteristics • Identify potential causes • Analyze root causes
5	Development and implementation of countermeasures	• Develop ideas for countermeasures • Define details of who, when and how for the countermeasure action plans • Implement the countermeasures
6	Verify the results	• Verify the results of the countermeasures • Compare with targets • Quantify results (tangible and intangible)
7	Standardize and control to sustain	• Develop or revise standards • Determine control methods • Communicate and educate stakeholders • Verify the results are being sustained

The Training Within Industry (TWI) programs that had a significant impact on raising U.S. wartime productivity were introduced to Japan in 1951 (Imai 2012). The TWI program was deployed by the U.S. government across 16,000 production facilities during the war. The survey results as of September 1945 from these plants found (Huntzinger 2008) the following:

▲ 55 percent had reduced scrap by 25 percent or more.
▲ 86 percent had increased production by 25 percent or more.
▲ 88 percent had reduced personnel by 25 percent or more.
▲ 100 percent had saved training time by 25 percent or more.
▲ 100 percent had reduced grievances by 25 percent or more.

A significant focus of the TWI programs was to raise the capabilities of supervisors through the three so-called J programs. The J programs addressed instruction skills, methods-improvement skills, and leadership skills. Designed as four-step processes, the J programs were and are today:

1. Job instruction (JI) to help supervisors train workers effectively
2. Job methods (JM) to "help the supervisors to produce greater quantities of quality products in less time, by making the best use of the manpower, machines and materials now available"
3. Job relations (JR) to help supervisors improve human relations.

Job methods in particular was a direct contributor to the *kaizen* approach, with its focus on breaking down the job, questioning every detail, developing a new method, and applying the new method (Table 2.3).

Far from being bound in the traditions and craftsmanship of Japanese culture, the analytical approach of *kaizen* came from JM, which was based on the practical, modern, Western scientific methods of the time.

Creative Idea Suggestion System

Leaders have called for improvement ideas in the form of suggestions from their people for hundreds of years, if not thousands. The first recorded suggestion system was seen in 1770, when the British Navy put in place a process to listen to individuals within the organization without risk of severe punishment (Robinson and Stern 1998). One hundred years later, companies in the United Kingdom and the United States were using suggestion boxes. The Ford Motor Company also adopted the suggestion

Table 2.3 Job Methods Training as the Basis for *Kaizen*

How to Improve Job Methods

A practical plan to help you produce greater quantities or quality products in less time, by making the best use of the Manpower, Machines, and Materials now available.

Step 1—Break down the job
1. List all the details of the job exactly as listed by the present method.
2. Be sure details include all:
   - Material handling
   - Machine work
   - Hand work

Step II—Question every detail
1. Use these types of questions:
   - Why is it necessary?
   - What is its purpose?
   - When should it be done?
   - Who is best qualified to do it?
   - How is the 'best way' to do it?
2. Also question the:
   - Materials, Machines, Equipment, Tools, Product Design, Layout, Workplace, Safety, Housekeeping.

Step III—Develop the new method
1. Eliminate unnecessary details.
2. Combine details when practical.
3. Rearrange for better sequence.
4. Simplify all necessary details:
   - Make the work easier and safer
   - Pre-position materials, tools, and equipment at the best places in the proper work area
   - Used gravity-feed hoppers and drop-delivery chutes
   - Let both hands do useful work
   - Use jigs and fixtures instead of hands for holding work
5. Work out your idea with others
6. Write up your proposed new method

Step IV—Apply the new method
1. Sell your proposal to the boss.
2. Sell the new method to the operators.
3. Get final approval of all concerned on safety, quality, quantity, cost.
4. Put the new method to work. Use it until a better way is developed.
5. Give credit where credit is due.

system, and it was from Ford that Toyota adapted the *kaizen* idea suggestion system in 1951. Remarkably, and unlike far too many companies that tried it across the United States and the United Kingdom, the creative idea suggestion system at Toyota has persisted for more than six decades.

Although it was largely a copy of the suggestion system at Ford, what Toyota did was to make it a requirement of managers to use the creative idea-suggestion system as a means to continually engage people in being mindful of the work they were performing. The suggestion system as it is run at Toyota and other leading companies is by no means the main or only way that *kaizen* is done. However, many of the elements that make a suggestion system successful are the same as those that make daily *kaizen* activity within natural teams successful, and such daily activity is the key to sustaining improvement over many decades. We will address this in detail in Chapter 6.

According to research by Miller (2007), a Japanese document that was distributed to Toyota employees in 1951 when the suggestion system was launched asked all workers for their ideas while preempting and answering such questions as:

▲ Why are we asking for your creative ideas?
▲ What type of ideas are we looking for?
▲ Who can submit ideas?
▲ How do we submit our ideas?
▲ How will the ideas be evaluated?
▲ What happens when ideas are accepted?

In answering the question, "Why are we asking for your creative ideas?" the document begins with a fairly typical statement that "progress has been made in reducing cost and improving quality, but we can still remove a lot of waste and improve quality." This is followed by the curious sentences: "If we look carefully, even in American automotive factories there is nothing remarkable about the production system. The accumulation of the joint efforts of every single person in the company is what makes them so productive." Toyota leadership presented the idea that the superior industrial productivity of the United States in 1951 was due to "the accumulation of the joint efforts of every single person" in U.S. companies rather than more advanced equipment, better management methods, or superior training, which were certainly also factors but not ones that the workforce could do anything about. This is the first key for successful suggestion systems, daily *kaizen*, and high engagement: focus people on improving their own work, on what is within their control.

Another sentence explains that the suggestion system will be different from the style of "idea festivals" held periodically in the past but will be an ongoing and widespread effort. There is an implied admission in the document that previous efforts were inconsistent. This section ends by stating several times very clearly that Toyota will pay for these ideas and will act quickly to implement them. Here Toyota made sure that *kaizen* was not just for show, that management and financial resources were committed to implementing ideas, and that the *kaizen* suggestion system would be "ongoing and widespread."

In answer to the question, "Who can submit ideas?" the Toyota document said that any employee can make suggestions, regardless of the type of work he or she does. In the beginning, managers were excluded. This initial exclusion may have served to roll the program out in manageable phases and to emphasize that the creative idea suggestion system was also a means of people development, where managers were responsible for coaching and implementing ideas of workers.

A very practical solution to a common challenge of suggestion systems is also given. In the early stages, certain people will gain more benefit from the fact that there are a large number of previously trapped, frozen, or ignored ideas and suggestions for improvement. Sometimes these are not *kaizen* ideas but rather people in a dysfunctional system finally being allowed to do their job normally. To avoid a sense of unfairness that one group of people is richly rewarded for picking low-hanging fruit when those people are really just doing their jobs normally, the document states that if the suggestion was one that was "normally expected in the course of work," then the idea would not be considered unless it was exceptional.

The answer to "How do we submit our ideas?"was very clearly, "Put it in the box," and specifies were provided as to where these suggestion boxes were located. Three other noteworthy points in this section were the statements, "If you can't write it yourself, it's okay to have a shop floor engineer help you" and even "We accept verbal suggestions" and "There is no need to gain your manager's permission before making a suggestion." In other words, there are no excuses for not participating. One key factor in the sustainability of *kaizen* at Toyota for over half a century is that the suggestion system evolved away from the box to the suggestion system to the process of today, whereby the area leader or supervisor discusses the documented *kaizen* idea and develops it fully with the worker who raised the idea.

Simple Yet Transformational Kaizen at Franciscan St. Francis Health

Joseph Swartz, Director of Business Transformation at Franciscan St. Francis Health in Indianapolis, Indiana learned firsthand the combined power of simplicity and employee engagement that comes from a thoughtfully designed *kaizen* suggestion system.

> The various healthcare organizations I had been working with in the past all had the desire and the need for continuous improvement, and had tried programs in the past, but those programs failed to empower the front-line staff. They did not give people the simple tools they needed to really make staff-led continuous improvement work for the long-term. When Franciscan St. Francis Health hired me seven years ago, it was initially to lead, organize, and execute lean six sigma projects. At that time they didn't know that they needed something like *kaizen*.
>
> After the first year I realized that with a staff of three facilitators executing 10 to 15 projects each year, we were only going to be able to engage 100 to 150 employees in personally experiencing what it takes to do continuous improvement. As the leader of continuous improvement for the organization, I knew we needed a way to involve all 4,000 of our employees. I was aware that our employees had to experience *kaizen* first hand by doing it themselves, and that they needed to do it many times repetitively before they really would "get it." I took the simple *kaizen* idea implementation model, trimmed down to 5 steps. Because of time constraints, I knew it had to be very simple at the beginning, implementable without much instruction.
>
> In 2007, Franciscan St. Francis Health CEO Bob Brody challenged all employees to participate in *kaizen*. In the first year we received 300 improvement ideas and in 2012 we received 4,000 *kaizen* ideas from employees. The result has been millions of dollars in savings, improved quality of healthcare, time and energy savings

for employees, all as a result of using people's talents. *Kaizen* is much more than changing things, it is about changing the behavior of other people, and this is very difficult. It was not always easy, and we had to learn what works by trial and error. This has been supported by our senior leaders, who ask employees about their *kaizen* efforts regularly. At an all-employee meeting our CEO Bob Brody will mention *kaizen* or lean six sigma half a dozen times during his presentation. It has been six years since we launched *kaizen* and it is still prominently in the messages from our leadership. When leadership keeps a consistent message of continuous improvement year after year, it builds confidence in the minds of employees that thinking about their work and giving their ideas really is an important activity for our organization and the patients we serve. Although we started small with simple *kaizen* suggestions, over time the *kaizen* ideas have become bigger and bigger.

Where Did the *Kaizen* Event Come From?

The *kaizen* event, in its popular five-day format, is an American innovation on a Japanese process. People are surprised to learn that Toyota and virtually no other Japanese companies that practice *kaizen* use the five-day *kaizen* event or *kaizen* blitz format popularized in the West. How can this be? In the late 1980s when Masaaki Imai and his consulting team were demonstrating the *kaizen* process to American companies, it was convenient to organize the sessions as "two-day *kaizens*," or intensive workshops (Imai 2012). These *kaizen* events were not intended as the one correct way of doing *kaizen* but as a demonstration of the impact of *kaizen*. "These were one-week marketing events," says Imai, recalling how *kaizen* events were introduced in the early 1980s. "After we had demonstrated the value of *kaizen*, we would go back and work with the management team to help them assess their current situation and develop a continuous improvement strategy. The *kaizen* events were never meant to be a consulting method, but due to the 'show' value, they became a very popular but not totally accurate way to implement *kaizen*."

At around the same time, the Japanese consulting firm Shingijutsu Company, Ltd., established by Taiichi Ohno's students from among suppliers in the Toyota Autonomous Study Group, began delivering these one-week *kaizen* events to companies outside Japan. The term *autonomous study* is a direct translation of the Japanese word *jishuken*, which is another word used within Toyota to mean "*kaizen* workshop." Often these are two- or three-day activities by a team of managers to deliver improvement through the application of a system such as *kanban* within a focused area. The goal was both to advance the Toyota Production System in a specific location and to provide opportunities for managers to expand their understanding through hands-on learning. The Shingijutsu consulting team was taught by Taiichi Ohno based on this type of *jishuken* approach. Although these types of workshops of less than one week are practical within Toyota suppliers based in the Nagoya area, they are not practical for consultants traveling to the United States from Japan, so the full-week *kaizen* event was born.

We can see from these examples that both the roots and more recent branches of *kaizen* are deeply American. We must hasten to add that the underlying concepts of customer focus, development of people, value and waste, and the scientific method are universal and neither American nor Japanese. What we saw in the best Japanese companies led us to believe that *kaizen* must be somehow unique to Japan because it was so foreign to our experience. If Japan is the soil in which the seeds of *kaizen* planted by Americans grew, we need to understand more about both the seeds—the basic values and assumptions—and the soil—the history and social context—in order to bring *kaizen* culture to our organizations.

Kaizen Is a Product of Japan's History, Not Its Culture

From a cultural standpoint, why did Toyota and a few other Japanese firms adopt the teachings of Deming and the TWI program successfully, whereas the United States and the West largely ignored these American treasures for half a century? Those who object to *kaizen* as fit for Japan but not for the culture in their own countries have built on a false premise that Japanese culture somehow makes the practice of *kaizen* easier to adopt.

Yet, the vast majority of Japanese organizations do not have *kaizen* cultures, do not have more than superficial knowledge of the Toyota Production System, and have cultures as functional or dysfunctional as one would expect from those in any industrialized nation. It is true that because of the postwar focus on quality and the rapid growth in the decades thereafter, paired with an unparalleled industrial policy of the government,

Japanese manufacturing became a powerhouse. However, this was due to urgent need, through herculean effort, and we could say even despite the limitations of the culture.

Because Japan is a high-context culture, communication in Japanese society is nonverbal and nonexplicit to a much greater extent than in low-context cultures such as the United States. This means that more is assumed to be understood within members of a group and less needs to be explicitly explained or communicated. When this works, it is a beautiful thing because fewer words are needed to communicate more owing to the presence of other contextual factors and shared understanding. People sharing the same language and similar backgrounds, religion, and levels of education will naturally have more unspoken understanding than the people of a diverse and spacious nation such as the United States.

However, a high-context culture is not necessarily adapted to success in complex manufacturing of technical products. Specifications must be written down, work instructions musts be detailed, and plans must be explicit and easily understood. These things are not natural and normal within Japanese communication and had to be learned with the help of quality gurus from low-context cultures. One such low-context, high-specificity cultural artifact and widely recognized improvement tool is the standardized work document set. Such a document specifies content, timing, sequence, and outcome for each step in terms of safety, quality, and the work to be done. One can only imagine how difficult it was for factory craftsmen within a high-context culture to accept this level of specificity.

Another example is the A3 document. Used to communicate problem solving activity, proposals, and even business plans, as well as strategy deployment documents, A3 documents are typically one- or two-page summaries on A3- or A4-sized sheets of paper. The characteristics of A3 documents are that they are brief, detailed, comprehensive, visual, and quite explicit. With an A3 problem solving document, the reader must be able to quickly and thoroughly understand the problem, the background situation, how it affects the customer, the target, the root causes, the countermeasures, how the problem solving process and results will be verified, and so forth. Little or nothing is left to the imagination. Managers within a *kaizen* culture must be skilled in coaching and guiding their people with this combination of brevity and detail. For a high-context culture such as Japan, the A3 document was a countermeasure to the problem that the national culture

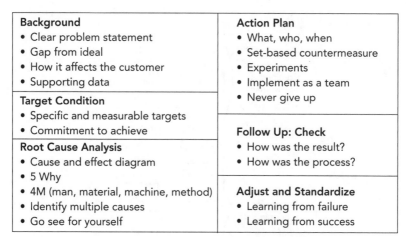

Background	Action Plan
• Clear problem statement • Gap from ideal • How it affects the customer • Supporting data	• What, who, when • Set-based countermeasure • Experiments • Implement as a team • Never give up
Target Condition • Specific and measurable targets • Commitment to achieve	
	Follow Up: Check • How was the result? • How was the process?
Root Cause Analysis • Cause and effect diagram • 5 Why • 4M (man, material, machine, method) • Identify multiple causes • Go see for yourself	**Adjust and Standardize** • Learning from failure • Learning from success

Figure 2.5 The A3 document as a low-context countermeasure for a high-context culture.

had a bias toward low levels of detail and the assumption that much was understood when little was said (Fig. 2.5).

Kaizen culture places a high priority on exposing problems, raising issues to management, experimenting, and learning through failure. Anyone who thinks that these are natural cultural characteristics has not spent enough time in Northeast Asian cultures. Exposing one's ignorance, challenging authority, and confronting failure are avoided in order to maintain harmony, a strong cultural value. The fear of "loss of face" in countries such as Japan, Korea, and China, along with a respect for authority and seniority, makes it difficult for most organizations to adopt the *kaizen* behaviors of exposing abnormalities, stopping and fixing, learning from failure, and even speaking openly about problems with those in authority. Equally, with effort and dedication, the workers of these countries are as capable as workers from any culture of adopting these practices. Taiichi Ohno wrote intimately and honestly about these and other early challenges of changing traditional mind-sets at Toyota into what we would recognize today as the *kaizen* mind-set (Ohno 2012).

Language is yet another barrier to creating a *kaizen* culture in Japan. So much is assumed to be understood with so few words. There are three alphabets and countless borrowed words from other languages, particularly in business and technical fields. This cultural artifact called language is rooted in a deep assumption among Japanese *kaizen* consultants who are

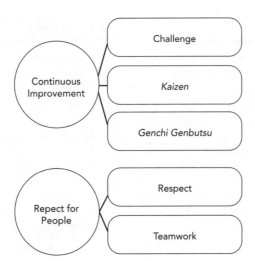

Figure 2.6 Nucleus of the Toyota Way.

not masters of English that somehow the Toyota Production System cannot work the same way beyond Japan because certain Japanese expressions and ideas simply cannot be expressed in other languages. Ignorance on this issue runs in both directions.

Recognizing that the traditional approach of mentor-mentee education in Toyota business practices was no longer adequate to keep up with the demand for rapid worldwide expansion of its operations, in 2001 Toyota codified key elements of its culture in what is called the *Toyota Way*. This included an explanation of the nucleus of continuous improvement and respect for people. An additional level of detail was provided to each of these two areas (Fig. 2.6).

It is revealing of the challenges faced with the globalization of operations that after 50 years of practicing *kaizen* and refining the Toyota Production System, with a bias against explicitly codifying and document-ing it, Chairman Fujio Cho felt that it was important to make these values, method, and principles explicit. In particular, the pairing of "continuous improvement and respect for people" is bold in its simplicity.

What's the Japanese for *Teamwork*?

Over the past 400 years, the Japanese have borrowed many words to express concepts in science, business, culinary arts, medicine, popular culture from other languages, including Portuguese, Dutch, German, French, and English, as well as from the Chinese language for much longer. When the Japanese wish to eat bread, they use Portuguese. When they ask for medical records, they use German. When pointing out a romantic pair, they use French.

There is little evidence of a tradition of competitive team sports in Japanese history prior to about 150 years ago when the Japanese began to be introduced from the West. Traditional sports such as sumo wrestling, archery, karate, kendo, judo, aikido, and various other martial arts exist, and although teams may compete against each other, these are individual sports with person-against-person not team-against-team matches. The one example of an old team sport is *kemari*, which dates back more than 1,400 years and is an import from China. This, however, is a *cooperative* game rather than a *competitive* sport, in which everyone tries to kick the ball and keep it up in the air for as long as possible.

There seems to be no tradition in Japan of groups of people practicing teamwork in the sense of competing team to team, with the exception of armed conflict, politics, commerce, and the occasional game of tug-of-war. It is no surprise, then, that the Japanese word for *teamwork* is *cheemuwaaku*, simply borrowed from English. Although there are words for *group action* and *helping out*, the Japanese have chosen *cheemuwaaku* as the best way to express teamwork. With a strong tradition of craftsmanship, industrial companies in Japan needed to convert from individual skill-based workshops to modern continuous-flow-oriented factories based on the model of Henry Ford's factories. This required a very different way of working together. Here again, as a high-context culture, the Japanese needed to borrow a special term to make explicit the way of working as a team. This should shatter the myth that the Japanese are culturally better team players than those of us from countries with longer traditions of playing soccer, baseball, or cricket.

A History of Adoption and Adaptation of Ideas

Why were TWI programs, the PDCA cycle, the suggestion system, scientific problem solving, and the teachings of Deming, Juran, Drucker, and others so readily received and developed into what we know as *kaizen* today in Japan but not in the United States? First, there is a millennium of history of adoption and adaptation in Japan that includes music, writing, food, religion, urban design, and many other cultural artifacts and beliefs from China. More recently, there have been periods of closing and opening to the outside world, during which the Japanese faced the shock of seeing the technological gaps, leading to rapid modernization efforts. These changes were met by resistance, and adoption of new ideas was not because of widely held cultural beliefs about adopting new things but out of the recognition and leadership of a few people who recognized that these powerful foreigners represented changes that could not be resisted and that society must adapt to survive.

There is a proverb, "No prophet is accepted in his own country," that underscores the fact that we are not able to listen to a tough message about change from one of our own people. Such a person is too familiar, and we do not respect his or her authority. For the same reason, consultants from Japan who teach *kaizen* are valued more than local consultants, and they are valued far more outside Japan than within.

Another factor is that we sense that they may be right, but change is inconvenient or not urgent. When things are going well, as was the post–World War II economy in the United States in the 1950s, there is even less incentive to listen to the voices of the prophets who demand change. At that time, Toyota went through bankruptcy and was forced to rely on banks to restructure. Indeed, without the special orders from the U.S. military during the Korean War creating enough demand to help Toyota grow its way out of the crisis, putting the company on a growth track far more quickly than domestic demand would have allowed, the history of Toyota, *kaizen*, and the Toyota Production System would look very different. It was a fortunate historical accident that this crisis was followed shortly after by exposure to so much valuable American management know-how.

A half-century after Toyota's bankruptcy, the company's main crisis now could be said to be the lack of a genuine sense of urgency. In order to prevent arrogance and self-satisfaction from setting in, especially among the younger generation of Toyota employees who may see themselves as elites who have joined the world's best manufacturer, Toyota leadership must renew and maintain a healthy dissatisfaction with the status quo, to the point of a sense of urgency. We can say that Toyota benefited from historical accidents that placed it on the course to growth, and the world has benefited from the resulting development of the *kaizen* method.

By the same token, just as Toyota management realizes that there is nothing inherent in the Japanese culture that makes the company more able to sustain its *kaizen* culture, we must realize that in order for us to succeed in emulating the shaping of an adaptive culture of excellence, we cannot make differences in national culture an excuse. Instead, we must find the sense of urgency, the inner fire to keep pursuing *kaizen*, regardless of the level of success we may achieve.

Scholar of probability, randomness, and uncertainty Nassim Taleb points out that it is precisely the ability to face uncertainty, accept risks, and engage in trial and error that is "America's asset." Taleb (2012) writes

> Like Britain in the Industrial Revolution, America's asset is, simply, risk taking and the use of optionality, this remarkable ability to engage in rational forms of trial and error, with no comparative shame in failing again, starting again, and repeating failure.

Taleb argues for *antifragility*, much more than merely adapting to change but growing stronger in response to shocks, through small, nimble systems that are adjusted via trial and error. It is encouraging for the future of *kaizen* culture in the United States that he observes that the willingness to take risk, fail, and persist trying is indeed an American cultural trait.

We need to concern ourselves less with the cultures of other organizations or countries and ask ourselves instead whether we have the *kaizen* mind-set and attitude that will allow us to succeed.

1. How great is the sense of urgency to change?
2. How strong is the attachment to the current ways?
3. How have we responded to change in the past?
4. How willing are we to humble ourselves to learn from others?
5. How prepared are we to fail, learn, fail, and learn yet again?

Kaizen must not be seen only as an instance of scientific problem solving, a method to capture small ideas bottom up, or a system to create and deploy performance-improvement plans top down. Excellence in creativity, innovation, and new-product development also emerge when *kaizen* values are adopted and put into practice. It is again only an accident of history that the most visible, tangible, and high-volume industrial processes were the first places to apply *kaizen*, a fact that has been rapidly changing over recent decades.

The False Dichotomy of *Kaizen* and Innovation

Innovation, although it can mean improvement to an existing product or process, is most often used by businesses to suggest revenue growth through new or improved products or services. A 2012 article in the *Wall Street Journal* showed that businesses seem to have innovation on their minds. In 2011, annual and quarterly reports mentioned innovation some 33,528 times, a 64 percent increase from 2006. The article argues that the word is being overused in an effort to signal to investors that companies will grow, but clearly, innovation is as old as humanity, and we must improve how we innovate. We have encountered resistance from time to time to the idea of *kaizen* as somehow stifling, uncreative, or killing initiative through regimentation. This is a false belief based on misunderstanding.

A *dichotomy* is a division or contrast between two things that are or are represented as being opposed or entirely different. People mistakenly think that *kaizen* means small, incremental improvement based on removing waste from an existing process, which somehow stifles innovation. Nineteenth-century French novelist Gustave Flaubert wrote, "Be regular and orderly in your life so that you may be violent and original in your work." Echoed in this advice is the belief that creativity is born from a foundation of stability, routine, and order, just as improvement ideas come from a foundation of standards. Whether we find as individuals that to live disordered Bohemian lives or disciplined mainstream lives enhances our creative output and self-actualization is a purely personal choice. However, in the context of organizations pursuing performance excellence, there is evidence that innovation is not stifled but enabled by the decisions people make within a *kaizen* culture.

What does it mean to apply core beliefs of *kaizen* to the business innovation process? How do successful innovators do it? The lean startup movement and the book by the same name (Ries 2011) describe how open-minded entrepreneurs are increasingly finding that *kaizen* leads to and enables innovation by adapting the core principles of the Toyota Production System, such as starting small and not overcommitting ahead of customer demand; pivoting, or taking customer feedback and changing often; pursuing root causes when solving problems; and learning quickly and often through small failed experiments. The crux of the lean startup approach seems to be failing fast and often in order to adapt, rather than failing big a few times in ways that may not be survivable for an early-stage venture. Information technologies in particular make it easier to create products, services, and business models with dramatically lower initial costs, but given their intangible nature, they also benefit from the lean practice of visual management of the development process and business planning. In fact, *kaizen* is more suited for startups than for traditional management precisely because *kaizen* is built on nearness to the customer, rapid experimentation, and seeking ideas from many people. As a consequence, quality, speed, and efficiency improve, but not at the cost of innovation, as some have argued.

Another common objection is that *kaizen*'s insistence on having standards inhibits creativity. It is not overstating the matter to say that without maintenance of standards and *kaizen* activity to improve the standards, innovation and development are not possible. This is a common misunderstanding—that standards are rigid, unchanging things—which they are not. Standards are ever-changing, simply the best known way *today*, which is only a reference point for future innovation. Taiicho Ohno famously said, "Where there is no standard, there can be no *kaizen*," and we can extend this further to say, "Where there is no *kaizen*, there can be no innovation." However, the integrity of *kaizen* and innovation is not about using lean tools or the *kaizen* method; rather, it is about the winning behaviors that emerge when people embrace *kaizen* values such as understanding the customer, having a secure environment in which to fail and learn, and acting with urgency, as we will see in Chapter 3. None less than the guru of innovation himself, Steve Jobs, said in an interview (*BusinessWeek*, 2004)

> Apple is a very disciplined company, and we have great processes. But that's not what it's about. Process makes you more efficient. . . . But innovation comes from people meeting up in the hallways or calling each other at 10:30 at night with a new idea, or because they realized something that shoots holes in how we've been thinking about a problem.

It is in this type of environment where people are encouraged to align toward solving shared problems, connect in formal and informal ways, and learn by respectfully challenging each other's ideas, all *kaizen* behaviors, that innovation flourishes.

In a typical startup, product-development group, or creative organization, senior management devotes far too much time to solving problems that result from lack of basic standards or lack of adherence to standards. This is problematic both because as time goes by people in the organization come to depend on these leaders as heroic problem solvers. Moreover, this type of

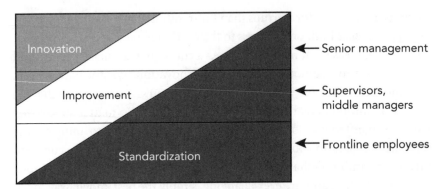

Figure 2.7 *Kaizen* flag.

front-line problem solving is not the highest and best use of the time and energy of the most senior leaders in an organization. Furthermore, frequently, the people who rise to become leaders of these organizations are themselves highly creative problem solvers who may be frustrated by the fact that they now spend all of their time managing rather than creating. Imai (1986) introduced this concept of the need for all levels to engage in maintenance of standards and the creative activity of improvement as the *kaizen flag* to explain the appropriate focus of attention by people within an organization on innovation, process improvement, and maintenance of standards (Fig. 2.7). However, this message was almost entirely missed by the readers of *Kaizen: The Key to Japan's Competitive Success,* perhaps because most of the cases studies and examples in that book involved standard-setting, maintenance, process-improvement, and *kaizen* projects in industrial settings.

What is worthy of note is the fact that although most of senior leadership's time is focused on innovation and improvement, a minor part of their focus remains on setting, checking on, and maintaining standards on the front lines. It is this balance of strategic, long-term, high-impact innovation with capability building through process improvement and vigilant attention to maintenance and upkeep of standards that yields sustainable competitive advantage.

Kaizen *in the Research and Development Team at Electrolux*

Introducing *kaizen* culture into a professional environment where, by the nature of knowledge work, the process is less visible and more reliant on individuals with strong beliefs in how the work should be done, such as integrating *kaizen* into a research and development (R&D) environment, brings a unique set of challenges. Tiziano Toschi, senior vice president of global R&D food preparation for Swedish household and professional appliance manufacturer Electrolux, explains:

> *Kaizen* isn't often considered as a tool for research and development activities, so it was big news for everybody when I introduced it. A lot of people had been exposed to *kaizen* in their work with manufacturing, but their reaction was, "What, you mean we have to apply that in our process? We don't manufacture."
>
> This is in some ways a strange reaction because R&D engineers are used to thinking that every product can be improved, and should be improved, and that's why they come to the office every morning. They know that next year they will come up with a product that is better than the one that they are launching this year.
>
> So my challenge to those people was, "Why can't you accept that the processes you use for debugging the software, creating a bill of materials, or generating a user manual can be improved?"
>
> The logic behind this question caused people to think about it, but the change in thinking did not come on its own. The average engineer is highly convinced that he knows better and you know less, so he is highly skeptical about any big news unless he invented it.
>
> It has to be an engineer to try to convince an engineer—it cannot be somebody else. And it's better that it is a peer rather than the boss—a peer has much more influence.
>
> What makes a real breakthrough possible, I think, is the mutual understanding that they can change their processes better than anybody else.
>
> The key at the beginning was our early adopters—people who were enthusiastic and ready to start with *kaizen* right away. We trained these people with the Kaizen Institute and then used them as internal trainers. Once we had a critical mass of trained people, this is where the snowball effect took over.
>
> Of course, there are always skeptics that are very hard to convince. Sometimes you have to isolate these people so [that]

they don't influence others, or even, in a worst case scenario, move them out of the company. To change their thinking, you have to show them so many winning situations that they can't deny the success. One or two won't do it.

Securing People to Develop New Products and New Customers Through Kaizen at Toyoda Gosei

Masao Nemoto led total quality control (TQC) activities within Toyota, its suppliers, and even the sales and distribution network for decades. Nemoto became president of Toyoda Gosei in 1982, during a period of slowing growth. He quickly realized that it was essential to shift human resources toward product development and sales. Although the manufacturing operations at Toyoda Gosei were already world class, the productivity of the white-collar processes was not at comparable levels. Nemoto's approach to deploying *kaizen* began by communicating the three aims and needs for improved productivity with leaders of the various departments (Nemoto 1983):

1. Productivity improvement was not merely to reduce costs in line with slow growth in demand, but to aggressively secure the human resources who can develop new products and develop new customer demand;
2. Therefore, the most capable people must be pulled out from their current responsibilities and others rotated into those roles; and
3. This freeing of these resources must be done based on deploying the concepts of Toyota Production System.

What does the "concepts of the Toyota Production System" mean within the context of freeing of these people from their current jobs into more innovation-oriented product development and new market development teams? With the target of improving productivity by 30 percent in 1½ years, Nemoto examined the white collar work to look for opportunities to eliminate, simplify, and combine tasks.

Following with the *kaizen* approach to create a sense of urgency, Nemoto did not wait until the 30 percent improvement was achieved to reallocate the best people from these departments but began by removing people first. This created both an immediate benefit by increasing the size of the innovation team and the drive to aggressively streamline the tasks and workload shared by the remaining staff. By communicating thoroughly the purpose and approach and by allowing a lot of freedom in how each department made changes, Nemoto was able to apply *kaizen* to quickly create additional capacity for innovation at Toyoda Gosei. We can infer from this example that the solid foundation of standards and process improvement at Toyota and its suppliers yielded consistent and strong financial results, created additional capacity, and allowed the company to invest in product development and launch truly innovative products, including the luxury Lexus line, the youth-oriented Scion brand, and the gas-electric hybrid Prius, as just a few visible examples.

The Scientific Method Within Organizational DNA

Many technological, managerial, and business model innovations are the application of certain scientific or economic principles to specific products and market environments. Lean manufacturing in general and, more specifically, the Toyota Production System were innovations that resulted from the repeated application of *kaizen* within the automotive supply-chain model over a period of decades. The principles of one-piece flow, built-in quality, downstream pull, and leveled scheduling and all the tools and techniques that enabled them were results of practical problem solving. Although in hindsight we can see that as a supply-chain model the Toyota Production System was the logical outcome of a resource-constrained organization taking the path of least resistance against the laws of queuing theory, fluid dynamics, system theory, and other laws governing the effective movement of material and information through the market, it was never an intentional design exercise to "innovate the delivery model."

Shimokawa and Fujimoto (2009) revealed that there was no master plan in the creation of the Toyota Production System, only innovations as a result of repeated experimentation. The stories of the early days of development of the Toyota Production System include a telling quote from Taiichi Ohno, the father of the Toyota Production System: "My first move as the manager of the machine shop was to introduce standardized work." Ohno's magnum opus was the development of the Toyota Production System, perhaps the greatest supply-chain innovation of the past few centuries, but his starting point was to set and maintain standards as the basis for improvement.

An in-depth academic study of Toyota's approach to management, covering more than 40 factories across the United States, Europe, and Japan over four years, was made by Bowen and Spear (1999). The study resolved the paradox of the apparently rigidly scripted Toyota Production System with its flexibility and adaptability to change. To this end, the authors proposed four rules that guide the design, operation, and improvement of every activity, connection, and pathway for every product and service at Toyota. The four rules are as follows:

1. All work shall be highly specified as to content, sequence, timing, and outcome.
2. Every customer-supplier connection must be direct, and there must be an unambiguous yes-or-no way to send requests and receive responses.
3. The pathway for every product and service must be simple and direct.
4. Any improvement must be made in accordance with the scientific method, under the guidance of a teacher, at the lowest possible level in the organization.

Furthermore, all rules must have built-in tests to signal abnormalities and deviations from the standard. This continual exposure and response to problems is what makes what appears to be a fragile and inflexible system quite flexible and adaptive.

To simplify further, we can say that *kaizen* cultures such as Toyota set standards for all processes, make deviations from standards visible, and educate and empower front-line personnel to improve based on the scientific method. The research of Bowen and Spear has helped to bring a level of academic respectability to the study of *kaizen* methods and the Toyota Production System. In addition, the explicit elevation of the scientific method as a DNA building block (rule 4 above) has made the Toyota Production System seem less foreign and has made it easier for highly educated professionals such as doctors, engineers, researchers, financial professionals, and educators to embrace the *kaizen* approach.

Like the DNA molecule, encoded in the simple word *kaizen* are the instructions to develop people and transform organizations. However, just as DNA must be understood in terms of how its components combine to create rules and build organisms, we must now unravel *kaizen* and study its building blocks, the essential core beliefs within *kaizen* culture.

CHAPTER 3

Core Beliefs Within
Kaizen Culture

Somebody once said that in looking for people to hire, you look for three qualities: integrity, intelligence, and energy. And if they don't have the first, the other two will kill you. You think about it; it's true. If you hire somebody without the first, you really want them to be dumb and lazy.

—WARREN BUFFETT

Creating a *kaizen* culture requires that we make an honest examination of our core beliefs and assumptions that result in group behaviors within our organization. Much of the literature and discussion of *kaizen*, lean, six sigma and operational excellence has been focused on its tangible aspects, the artifacts and behaviors. While the use of tools and artifacts as routines to build habits in what Rother (2010) calls *kata* does bring about behavior changes, sustained modification of group behaviors is only possible when it is anchored in the deepest level of culture, the core beliefs (Table 3.1). Only by understanding the current state and comparing it against our desired target condition will we be able to build a plan to put a *kaizen* culture into practice.

This initial grasping of the current state of our core beliefs is in itself no small achievement. The question we must ask ourselves before we begin is, "Why don't more lean, six sigma, and business efforts pay greater attention to the deeper part of organizational culture, people's core beliefs and assumptions?" First, these most often begin as financial results-driven efforts and only become interested in values when it becomes clear that there is a cause-and-effect chain between core beliefs, behaviors, and sustainability of improvement results. But even this has become an increasingly mainstream topic with many avenues for advice. A secondary factor is that people confuse values, principles, philosophies, mission, purpose, and so forth. Beliefs are those things we hold to be true, on which we place great personal importance, even passion. We must feel strongly for these things to be core beliefs, with our hearts and not just our heads. Third, too often the work of culture and core beliefs is delegated by the senior leadership team to someone perceived to have relevant expertise, often in human resources or organizational development areas. Although functional expertise and leadership are important, without direct and strong links to the front lines, familiarity with the artifacts of the current culture, and an ability to direct both the eyes and the feet of senior leadership to see for themselves, the discussion of core beliefs cannot be genuine, and becomes something soft and abstract. But even before we point our feet to the direction of the *gemba*, we must know both what to look for and what our eyes are telling us. In this chapter we will attempt to set the foundation for leaders to develop a correct understanding of the core beliefs of *kaizen*, how they are put into practice, and how they bring about a high-performance culture.

Integrity, Core Beliefs, and the Act of *Kaizen*

Very simply put, the core beliefs underpinning *kaizen* are the inner motivations and direction for doing the right thing. The motivation in a specific situation may stem from an inborn sense of human fairness, organizational doctrine that has been drilled into the individual, or a personal creed developed over years of experience. In most of us, they are a combination of all three. When a person consistently acts in harmony with a set of core beliefs that are shared or assumed to be shared by the individual and others with whom that individual interact, we say that such a person has integrity. A basic example is the low ethical worth most societies place on lying, stealing, or cheating. People who do these things have low integrity. At the same time, people who fail to keep their commitments to do good things, even when they are not actively doing bad things, are also said to have low integrity. Integrity itself is a *kaizen* core belief because we find worth not only in the result of people's actions but in the process or effort put forth to achieve the promised result.

Warren Buffett's insight into the primacy of integrity in selecting individual employees can be extended to how people are inducted, trained, and coached after they are hired. Buffett alludes to the fact that the most intelligent, well-intentioned, and energetic people can bring about disastrous results by acting without integrity. However, it is also true that such people can act with high integrity, keeping their word and acting ethically, but following false principles, unsuccessful business models, or beliefs that are not true. To some degree, we can excuse this as ignorance that is removed through education, training, and mentoring. The organizational culture must be based on core beliefs that stand up in practice and deliver results, not only those that feel good or that have a long tradition. This is why respect for people and the scientific method are integral to the value set within any organizational culture. These core beliefs are self-correcting and have stood the test of time.

Warren Buffett also said, "You're neither right nor wrong because other people agree with you. You're right because your facts are right and your reasoning is right—and that's the only thing that makes you right. And if your facts and reasoning are right, you don't have to worry about anybody else." Although his authority is impressive, we must not take one of the five wealthiest men in the world at his word; each of us must decide whether integrity is primary, which values to espouse, and how to adapt without compromise when our beliefs prove untrue in practice. We propose that the set of *kaizen* core beliefs introduced in this chapter provides such a roadmap.

Guiding Principles Drive Lean Six Sigma Deployment at Jabil Circuits

Jabil Circuits, Inc. is a global EMS (electronics manufacturing services) company based in St. Petersburg, Florida. Since 2005, Jabil has been on a long and ambitious journey of cultural transformation. With aspirations of becoming a company of "165,000+ problem solvers," Jabil uses *kaizen* in many ways to develop everyone into a problem solver "focused on delivering only value to our customers." This is strongly supported by top management and orchestrated by internal consultants and trainers in lean six sigma.

"We are not promoting a lean six sigma program. We are transforming the company by instilling a mindset and culture of continuous improvement that is evident in everything we do," said Mike Matthes, Senior Vice President,

Worldwide Operations. Since 2011, nearly 10 percent of the workforce has been trained in lean six sigma methodologies and more than 30,000 *kaizen* projects of various sizes were completed by employees. As a strongly results-oriented organization, the tangible improvement results were a great source of additional momentum, and the business leaders demanded the rapid expansion of the training and projects across all 60+ sites worldwide.

However, those closest to the front lines of the transformation were beginning to see the limitations of the tool-focused and results-driven approach to continuous improvement. The tools that were intended to solve problems and problem solvers were applied in some cases without much thought, only to fulfill a requirement to "do lean six sigma." As a counter-measure, Jabil developed a set of guiding principles to remind people of the intent and purpose of the tools. Jabil's lean six sigma guiding principles are

▲ Customer satisfaction
  ▼ Create value for our customers
▲ Velocity
  ▼ Produce only what is needed, when it is needed, in the right amount
  ▼ Eliminate anything that stops the flow of value creation
  ▼ Focus on value streams
▲ Build-in quality
  ▼ Never pass a defect on to the next process
  ▼ Make problems visible
▲ Continuous improvement
  ▼ Relentlessly eliminate waste
  ▼ Embrace scientific problem solving
  ▼ Observe problems first-hand
▲ Cultural enablers
  ▼ Develop people
  ▼ Promote teamwork
  ▼ Lead with humility

"At first there was resistance to putting emphasis on these guiding principles," said Jaime Villafuerte, Director of Lean Six Sigma for Jabil. "Many of the leaders within operations as well as engineering have strong technical backgrounds, and perhaps these principles seemed too fuzzy and abstract to them. Today these principles are increasingly becoming core beliefs that drive behaviors and cultural transformation at Jabil."

Table 3.1 Core Beliefs as Foundation of *Kaizen* Culture

Kaizen Core Beliefs	Toyoda Precepts	Nemoto's Management Creed	Zappos Family Core Values	Imai's 10 Rules for Kaizen	14 Principles of the Toyota Way	Deming's 14 Points for Management
			Strive for virtue, to be good			
Humility	Be reverent, and show gratitude for things great and small in thought and deed	Top management must ask "What can I do to help?" during audits	Be humble	Discard conventional rigid thinking about production Think of how to do it, not why it cannot be done	Become a learning organization through relentless reflection and continuous improvement	Adopt the new philosophy
Alignment			Be passionate and determined	Remember the opportunities for kaizen are infinite	Base your management decisions on a long-term philosophy, even at the expense of short-term financial goals	Create constancy of purpose toward improvement of product and service, with the aim to become competitive, stay in business and to provide jobs Put everybody in the company to work to accomplish the transformation

(continued on next page)

Table 3.1 Core Beliefs as Foundation of *Kaizen* Culture (Continued)

Kaizen Core Beliefs	Toyoda Precepts	Nemoto's Management Creed	Zappos Family Core Values	Imai's 10 Rules for Kaizen	14 Principles of the Toyota Way	Deming's 14 Points for Management
Security			Build open and honest relationships with communication		Use visual controls so no problems are hidden Standardized tasks and processes are the foundation for continuous improvement and employee empowerment	End the practice of awarding business on the basis of a price tag Drive out fear, so that everyone may work effectively for the company
Service	Be at the vanguard of the times through endless creativity, inquisitiveness and pursuit of improvement	Give up your best person when doing job rotation	Deliver WOW through service		Use 'pull' systems to avoid overproduction	

Respect	Be kind and generous, strive to create a warm, homelike atmosphere	When the learner does not understand, try another way of teaching; The rehearsal is an ideal place for training; Everyone speaks up; Do not scold	Pursue growth and learning; Create fun and a little weirdness; Build a positive team and family spirit	Develop exceptional people and teams who follow your company's philosophy; Grow leaders who thoroughly understand the work, live the philosophy, and teach it to others	Institute training on the job; Remove barriers [...] to pride of workmanship; Institute a vigorous program of education and self-improvement
Process		Wisdom is brought out when faced with hardship; Do not make excuses. Start by questioning current practices	Do more with less	Use only reliable, thoroughly tested technology; Level out the workload; Go and see for yourself to thoroughly understand the situation	Institute leadership. The aim of supervision should be to help people and machines [...] do a better job

(continued on next page)

Table 3.1 Core Beliefs as Foundation of *Kaizen* Culture (*Continued*)

Kaizen Core Beliefs	Toyoda Precepts	Nemoto's Management Creed	Zappos Family Core Values	Imai's 10 Rules for Kaizen	14 Principles of the Toyota Way	Deming's 14 Points for Management
Urgency	Be practical and avoid frivolity	*Kaizen*, and *kaizen* again If my orders don't have due dates, ignore them Company audits that result in no top management action are useless audits	Embrace and drive change Be adventurous, creative, and open-minded	Ask "Why?" 5 times and seek the root cause Correct mistakes at once. Do not seek perfection. Do it right away even if for only 50 percent of target Do not spend money for *kaizen*	Build a culture of stopping to fix problems, to get quality right the first time	Improve constantly and forever the system [...] of production and service, to improve quality and productivity, and thus constantly decrease costs Cease dependence on inspection to achieve quality
Connection		Coordination between departments is an essential skill of managers			Create a continuous process flow to bring problems to the surface	Break down barriers between departments

Consensus	Seek the wisdom of ten people rather than the knowledge of one	Make decisions slowly by consensus, thoroughly considering all options; implement decisions rapidly	Eliminate slogans, exhortations, and targets for the work force Eliminate work standards (quotas) on the factory floor Substitute with leadership
Sharing	Be contributive to the development and welfare of the country by working together	Respect your extended network of partners and suppliers by challenging them and helping them improve.	

Precepts, Points, Creeds, and Principles

Scholars and management gurus who have studied and attempted to guide leading organizations have directed us to look beyond the superficial methods, tools, and systems to the underlying principles. Deming had his 14 points for management (Deming 2000), which were hugely influential on Japanese management but sadly less so elsewhere. Liker (2003) identified the 14 principles of the Toyota Way mapped against a four-level pyramid with philosophy (long-term thinking) at the base; process (eliminate waste), people, and partners (respect, challenge, and grow them) in the middle; and problem solving (continuous improvement and learning) at the top. A comparison of these and other precepts, values, and creeds introduced in this book reveals the common themes (Table 3.1). The commonalities are due to the ease of observation of the artifacts or behaviors that emerge from the cultural assumptions or beliefs (Fig. 3.1).

An example is the belief that thoroughly understanding a process and taking corrective action at the root cause level are better than quickly jumping to solutions, as evidenced by the behavior of the *andon* signal, fishbone diagrams, and the five-why process in workplaces to facilitate this

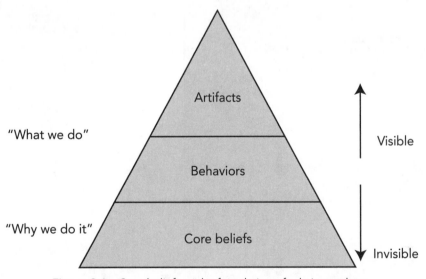

Figure 3.1   Core beliefs as the foundation of a *kaizen* culture.

root cause analysis. Another factor for the commonality is no doubt mutual influence among these ideas as the older points, precepts, and creeds are studied and adapted. However, we believe that there are timeless, universal principles at work that are rooted in humanity, psychology, behavioral economics, and even simple logic. We have attempted to identify the root-level assumptions that have generated the many precepts and principles, calling them *kaizen* core beliefs. But first we will review an example of how these principles develop into deep cultural assumptions.

Toyoda Precepts and Their Impact on Toyota's Culture

A charismatic founder's vision and values have a great impact on the development of a country, society, or organizational culture. Introduced in 1935 by founder Sakichi Toyoda, the Toyoda Precepts were words that he lived by and remain carved on both literal and figurative pillars at Toyota. These precepts espoused gratitude, service, development of self and others, and hard work.

1. Be contributive to the development and welfare of the country by working together, regardless of position, in faithfully fulfilling your duties.
2. Be at the vanguard of the times through endless creativity, inquisitiveness, and pursuit of improvement.
3. Be practical, and avoid frivolity.
4. Be kind and generous, and strive to create a warm, homelike atmosphere.
5. Be reverent, and show gratitude for things great and small in thought and deed.

Hino (2005) added insight by making modern interpretations of these precepts that were written in a very different time. To reword and summarize Hino, these five precepts call on all employees to join together to improve the broader world, learn from the past and strive to create cutting-edge ideas, remove what is needless and focus on what is truly effective, help each other in a family atmosphere with a shared purpose, and give back to society and the earth itself with an awareness of the blessings we receive. What is important to today's Toyota employee is what Deming would call "constancy of purpose"—the steadiness of beliefs and values over time. This constancy helps people to make good decisions, those that are aligned with the shared purpose of the organization. This has been the impact of Toyoda Precepts on the company's culture over nearly a century and on hundreds of thousands of people worldwide.

The Management Creed of a Toyota Executive

Encouraged by the subordinates whom he always coached and advised, Nemoto (1983) wrote down the 10 aphorisms that became his creed for management over four decades at Toyota. We can see the inheritance of the Toyoda Precepts within these 10 statements. They contain many enlightening points when viewed from the perspective of assumptions and beliefs that shape the behavior of leaders and ultimately the organizational culture. They are

1. *Kaizen,* and *kaizen* again.
2. Coordination between departments is an essential skill of managers.
3. Everyone speaks up.
4. There is a reason I do not scold.
5. When the learner does not understand, try another way of teaching.
6. Give up your best person when doing job rotation.
7. If my orders don't have due dates, ignore them.
8. The rehearsal is an ideal place for training.
9. Company audits that result in no top-management action are useless audits.
10. Top management must ask, "What can I do to help?" during audits.

Many of these are self-explanatory. A few of them invite additional commentary.

The second principle concerns what in today's popular parlance may be termed *cross-functional management* or *end-to-end thinking* or even *value stream management.* Nemoto states the danger inherent in an increase in functional silos within organizations. He lists several essential actions to enable cross-functional management. The first is for top management to communicate to department managers that they value coordination and cooperation between departments, going so far as to state, "Managers who cannot do this are not qualified to lead their departments." Second, management must set up programs to develop these skills within the department and section managers, with the direct involvement of senior leaders via on-the-job training. Finally, the department manager who is the type of person who develops personality conflicts or is unable to work with certain other department managers is also disqualified from such leadership roles within a *kaizen* culture.

It is important not to misunderstand the third principle as merely giving everyone a chance to say whatever is on his or her mind. It is not an invitation to useless talk, talk without action. The meaning of "everyone speaks up" is to align the people within the organization toward goals and actions by hearing all voices and opinions in the course of developing the goals and actions. At the specific venues for planning or review, every person is asked to speak. This encourages debate in order to develop ideas into concrete and executable plans by using the wisdom of many rather than the expertise of one.

Within large organizations, it is common to see an unreasonable, unattainable volume of top-down mandates being piled onto subordinate managers via one-way communication. When each person is called on by name and the leader listens to what each person has to say, silent disagreement becomes much less likely, and engagement increases. The traditional senior leader who relies more on his or her own authority and knowledge often signals through his or her speech and action the message that "I want to hear the sound of my own voice." When the leader says, "I want to hear from everyone," and follows through by calling on each person by name, listening, and summarizing the result of collective input, the organization is able to build on the experience and wisdom of many and become more resilient and adaptive rather than relying on the opinion of one.

Nemoto explains that he never scolds or imposes penalties on his subordinates because he does not want people to hide problems but instead to talk openly about their mistakes so that they can learn from them. The price of not being scolded is that Nemoto was a demanding teacher and kept his subordinates to high expectations. This approach of avoiding, quite in contrast to the scolding given by Taichi Ohno and others, reveals once again that there is no single Japanese style that is central to the *kaizen* culture, only common assumptions. Although differing in personal styles, both Ohno and Nemoto attempted to develop people who maintained high standards, readily exposed problems, and developed management skills such as problem solving through practical application.

Nemoto's sixth principle is based on the practice of periodic job rotation within organizations that is common in Japan and to a lesser degree in other countries. Traditionally, in Japan, young people apply to and join a company, whose human resources department then assigns them to a position based on aptitude. Job rotation is a method of broadening and deepening the aptitude and fit of the employee for the company in the long term. In countries outside Japan, it is typical for a young person to apply for a specific job based on his or her education and training, such as software development, marketing, or finance. It would seem strange or even threatening to be rotated out of this field into an entirely different one, such as from software engineering into marketing. Management development programs in many multinational companies will rotate high-potential employees on a two- to three-year basis, but this is often based on individual development plans rather than on the company goals for cross-training across departments and functions. Both approaches aim to develop and retain talent. Nemoto's philosophy was to offer up the best talent to leave his department and rotate to another team in order to provide maximum benefit to the receiving end and the company as a whole. There is an implicit assumption here that interdepartmental cooperation exists and that what is good for the whole company is good for one's own department.

In the eighth principle, "The rehearsal is an ideal place for training," Nemoto is referring to rehearsals or practice sessions for quality circle (QC) presentations, policy deployment reviews, or top-management audits of the factory. There is an assumption that people will practice their presentations before giving them to top management and that middle management will listen and coach during these rehearsals. One of the deep assumptions within a *kaizen* culture is that a key responsibility of management is to develop the people within their organization. The act of giving a presentation brings together management skills, including verbal and written communication, time management, ability to organize information, and logical thinking. The rehearsal is the ideal place to observe, coach, and develop these skills.

Each of these principles suggests deep cultural assumptions of the organization that define a *kaizen* culture. However, it is important to remember that Nemoto had to specifically and explicitly teach these principles, often writing them down and reviewing them with his subordinates, because these are not natural Japanese cultural values. These were not even explicitly defined Toyota values at first. Toyota people spoke of these ideas as being "in the air" people breathed at Toyota, as if cultural values were transmitted by scent. These principles emerged through experience of leaders like Nemoto and gradually became part of the culture taught from person to person over many years as they were reinforced by the artifacts, the physical systems that resulted from the creed. Nemoto's management creed, like that of many other leaders who grew up at Toyota, was only informally documented, his own being shared through his book only years after his retirement. Many modern organizations are taking a much more dynamic and active stance in identifying, communicating, and perpetuating the core beliefs of the culture across the organization.

Explicitly Defining Core Values at Zappos

Although they are the most influential aspects of culture, one's beliefs and deeply-held assumptions are invisible. Therefore it is important to make these explicit and visible if we wish to put them into practice. In a sense, these are ideals that we strive toward but may fall short of embodying on a day-to-day basis. The actualization of core values into practice requires daily management practices and checking and coaching by managers who care about culture. One company has taken the explicit definition and reinforcement of culture to a higher level. Zappos is a highly customer-oriented online shoe and apparel retailer that has grown rapidly, topping $1 billion in sales in the 15 years since its founding in 1999. The company's success has come in part from the loyalty and relationships built with its customers. The primary sources of the company's rapid growth have been repeat customers and recommendations of customers, with 75 percent being repeat buyers. In addition to the renowned customer-service model, Zappos has placed a high value on *kaizen* and an explicit focus on nurturing its culture. Each year Zappos publishes a 480-page *Culture Book* (Fig. 3.2).

The Zappos company website (www.zappos.com) explains its unique culture, describing in detail the Zappos "Family Core Values":

> As we grow as a company, it has become more and more important to explicitly define the core values from which we

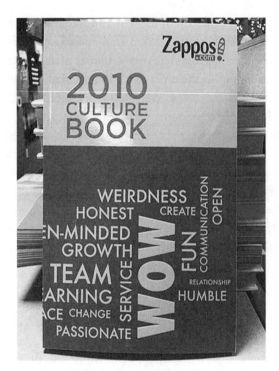

Figure 3.2 Zappos' *2010 Culture Book.*
(Wikimedia Commons.)

develop our culture, our brand, and our business strategies. These are the ten core values that we live by:

1. Deliver WOW Through Service
2. Embrace and Drive Change
3. Create Fun and a Little Weirdness
4. Be Adventurous, Creative, and Open-Minded
5. Pursue Growth and Learning
6. Build Open and Honest Relationships with Communication
7. Build a Positive Team and Family Spirit
8. Do More with Less
9. Be Passionate and Determined
10. Be Humble

At Zappos, customer service is famously not a department but the entire company. The company has learned that the right process yields the right results. That is, the right process for shaping and strengthening the organizational culture yields excellence in customer service and, with it, rapid growth and profits. Zappos CEO Tony Hsieh shared his management philosophy in an interview (Cerny 2009):

> Many companies think only one quarter ahead or one year ahead. We like to think about what we want our brand and culture to be like 10 or even 20 years down the line. In general, with a 10- to 20-year timeline versus a three- to five-year timeline, relationships are much more important. What you do after taking someone's money, such as customer service, matters much more than what you do to get their money, such as marketing.

For too many corporations and their leaders, the relationship with the customer quickly stops once the profit has been reaped. Futurist and author Alvin Toffler said, "Profits, like sausages . . . are esteemed most by those who know least about what goes into them." In contrast, the long-term thinking of Hsieh is another key cultural value, one that must be lived by the leadership in order to prepare for and adapt to changing customer needs, technologies, and market conditions. This level of attention to detail to employee behaviors and the front-line processes that create and maintain value for the customer is also a deep assumption of *kaizen* cultures. In the case of Zappos, the customer service, or postsale customer experience, is the company's front line, its *gemba*.

From Core Beliefs to Shared Purpose

We have identified a set of instructions at the root of these various principles, precepts, creeds, and management philosophies that we call the *core beliefs of kaizen culture* (Fig. 3.3). They are the beliefs that good things happen when we:

1. Cultivate humility and open-minded curiosity
2. Align the team toward a shared long-term purpose
3. Establish a secure, safe, stable, blame-free environment
4. Respect individuals and teams, nurturing human potential
5. Understand the purpose of your work, who you serve and what these people value

6. Understand your processes scientifically and through direct observation

7. Act with urgency to correct even the smallest faults or make the smallest improvements

8. Connect people, information, materials, and processes to improve the overall system

9. Build the consensus needed to execute with speed and certainty

10. Share with the community what is good about your organization with passion

We have arrived at these *kaizen* core beliefs by observing how leading organizations that we have studied or have worked with are putting their beliefs into daily practice. We share in this book examples of these core beliefs helping to create *kaizen* cultures, from Toyota, Zappos, Virginia Mason Medical Center, Jabil, and many others. Beliefs in alignment toward long-term purpose, understating of one's processes, building consensus, and humility are not always easy to see in practice when walking through a world-class organization. However, on interacting with the people in such organizations, these qualities can often be easily sensed. We believe that although not every organization that embraces "security" as a core *kaizen* beliefs will use *andons* and visual controls to make problems visible, no organization attempting to use these tools will succeed unless it is built on the foundation of these values. At the base of these core beliefs is one very important belief and motivator: virtue. Unless we have a desire to "be good," none of these *kaizen* core beliefs will have worth to us, or worse, they will do harm and no longer be *kaizen* by our definition.

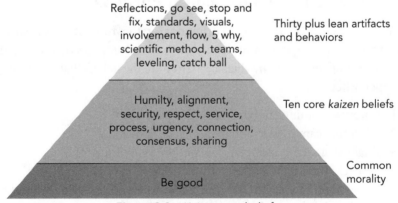

Figure 3.3 *Kaizen* core beliefs.

Throughout this book, we will illustrate the link between *kaizen* core beliefs and the various artifacts and tools that can be observed in organizations practicing *kaizen*, total quality management (TQM), lean management, and business excellence. We will attempt to make the case that beginning from these core beliefs of *kaizen* culture, and possibly others, a committed leadership team can succeed in building an organization that is adaptive, excellent, and sustainable. It may be useful to think of these core beliefs as ideals to strive toward or guideposts on our journey, with the associated artifacts and tools (Table 3.2) being countermeasures to close gaps in our progress.

Table 3.2 Core Beliefs and the Emerging Artifacts and Behaviors of *Kaizen* Culture

Kaizen Core Beliefs	Description	Emerging Artifacts and Behaviors
Humility	Humble open-mindedness; willingness to be wrong in order to learn better ways	Servant leadership; letting go of paradigms; learning from other organizations; not making excuses; willingness to experiment; reflection
Alignment	Aligning people towards a shared long-term purpose that is more financial	Passionate commitment to continuous improvement and service excellence; long-term investment in people and community; total engagement in the transformation; *hoshin kanri*
Security	Physical, emotional and professional safety; stability of the business; blame-free environment; safe to expose problems; attaching process faults but not people; safe to try and fail	Safety first practices; stop and fix; visual controls; 5S; no layoff policy for *kaizen*; *andon* system; two-way communication; standards
Respect	Respect for humanity and the individual	Development of people; leaders as teachers; job rotation; cross training; total engagement; family atmosphere; fun workplace
Service	Understand who it is that you serve and what they value; internal customer; external customer	Customer-aligned organizational structures; pull signal; value stream design; voice of customer

(continued on next page)

Table 3.2 Core Beliefs and the Emerging Artifacts
and Behaviors of *Kaizen* Culture (*Continued*)

Kaizen Core Beliefs	Description	Emerging Artifacts and Behaviors
Process	Understand your processes for designing the work, doing the work, improving the work	Value stream mapping; go see/ *genchi genbutsu*; low-cost automation; level workload; process-aligned frontline management; do more with less; standard work for leaders; problem solving process
Urgency	Act with urgency to correct faults and improve, no matter how small the opportunity	Dissatisfaction with status quo; stop and fix; 5 why; *kaizen*; improve constantly
Connection	Connect people, processes, information and material movement in order to break down barriers, expose problems and serve customers better	Value stream thinking; cross-functional teams; flow; visual management
Consensus	Make decisions through consensus, taking enough time to ensure quick and certain implementation	Catch ball; everyone speaks; team-based *kaizen* activity; daily shift start meetings
Sharing	Share best practices; teaching *kaizen* to the community outside of work, giving back to the community, sharing whatever is both good and gives you passion	*Yokoten (best practice sharing)*; benchmark tours; customer-supplier collaboration; volunteer *kaizen* at local non-profits

Some of the behaviors emerge out of a combination or interaction of several core beliefs. For example, the notions of respect for the individual, security and stability, and understanding the customer and the process result in the need for a leveled workload to reduce strain and overburden on the people and process. How these deeper beliefs are realized in practice will differ based on internal and external factors affecting each organization. What humility and open-minded curiosity means to a young startup will likely be very different from what it means for a large, old, well-established multinational corporation. Whereas the startup may already have this as a strong shared assumption and need little effort to translate it into action, a more complex organization will need to practice a lot and reinforce the message in different ways. The following examples illustrate ways in which many of these *kaizen* core beliefs are being put into practice in the world.

Alignment with Long-Term Purpose at Medtronic

At Medtronic's Jacksonville, Florida, plant, you won't find performance ratings on a spreadsheet—they are posted throughout the plant for all to see. This simple and surprisingly open visual approach to information flow has been extremely successful both financially and in building an engaged workforce. Here Emmanuel Dujarric, senior director of Medtronic Surgical Technologies, explains how *kaizen* values are lived at Medtronic:

> Our plant in Jacksonville, Florida, employs approximately 500 in the manufacture of medical devices. Our approach to *kaizen*, which we introduced in 2003, is to keep things very simple and very visible so [that] everybody can understand them. We have replaced most electronic information with visual information, to the point where our plant could now run without computers. Our approach has been, "Don't give me a big spreadsheet. We'll just say that if it's bad, it's red; if it's in trouble, it's yellow; and if it's okay, it's green. I don't need to know anything else."
>
> We manage at all levels by those three colors, and instead of entering the information into a system, we post everything on the walls. If it's green, we don't talk about it. The reds and yellows— the abnormalities—are what we're concerned with. So our walls are covered with entries of how we go about solving each problem. That's what managers of any operating area need to pay attention to. If an order wasn't delivered, or there is a quality issue or a staffing issue, or we are not meeting our safety target, we make all these visible.
>
> So when visitors come and see all the problems on the walls, they say, "How come you have so many problems?" We answer that we are problem fixers and problem preventers, and the wall shows us what we need to do. The other option would be to put everything on a spreadsheet that nobody looks at.

Every six months, we align our targets according to the direction of the corporation and post them. We do this for every function, including the HR and finance departments, and the process that guides the direction of the company. My personal performance goals are posted as well. Of course, the finance people are pretty happy with our results, or we wouldn't still be doing this.

We've also made our work order process completely visual by moving several manufacturing lines into the distribution center. There, we have shelves full of finished goods. When an order is shipped to a customer, this creates a visual signal, and since the line is close by, people immediately know what they need to build. There is nobody sitting at a desk analyzing demand and creating work orders—that function no longer exists.

The other side of this is that we challenge all of our employees to take part in *kaizens* on a regular basis. There's no bureaucracy around this, and *kaizens* are conducted without any management oversight. Any employee has the ability and the authority to stop working and make an improvement on the floor whenever they feel they have to do it. Then, every Friday, we get all the employees in the room, and they have the opportunity to present what improvement they have done in the last week. We get about 500 *kaizens* every quarter.

We don't rely on classroom training for *kaizens*. People learn by doing, and if somebody needs help, they can grab a couple of other people who know how to use the tools. Of course, people make mistakes, but we encourage them to learn from that and move on. Some of our people are becoming excellent running *kaizens* and are climbing the ranks of our company.

We have also eliminated titles when we are doing *kaizens* — it doesn't matter if you are an employee in the line or a director. And when we look back on results, it's been proven over and over again that the biggest breakthroughs we have had at Jacksonville came from line employees. People who work in manufacturing know about waste, and they know what improvements can do. As managers, we have to remain humble. In fact, we consider it our job—our only job—to serve the workers on the line.

This is an excellent illustration of how Medtronic has found a recipe for success in aligning people with the goals of the organization, making it safe and easy to expose problems, giving people support, connecting processes and removing barriers between people, giving people opportunities and the needed respect to make the improvements necessary to meet the challenges, and allowing people to attain a sense of satisfaction and engagement.

The Trust and Security to "Stop and Fix" at NUMMI

John Shook, chairman and CEO of the Lean Enterprise Institute and author of multiple books on lean management, was hired in 1983 by Toyota to develop and deliver training for the New United Motor Manufacturing, Inc. (NUMMI), joint venture with General Motors (GM). Recounting his experience in helping the American management team to adopt this behavior, Shook (2010) writes

> When NUMMI was being formed, though, some of our GM colleagues questioned the wisdom of trying to install *andon* there. "You intend to give these workers the right to stop the line?" they asked. Toyota's answer: "No, we intend to give them the obligation to stop it—whenever they find a problem."

To the people from GM, the *andon* represented a transfer of power or authority from management to the workforce. Based on past labor-management relations, it probably seemed like a bad idea to give their adversary more control over the production line. To the Toyota people, the obligation of the worker to stop work to expose a problem and call for help was a basic assumption, an assignment of responsibility built on trust and respect. This one episode demonstrates the large gap in expectations that often exists when adopting any one of the artifacts of a *kaizen* culture, namely the tools of lean management such as the *andon* system. It is safest to expect and plan for these types of encounters during a *kaizen* transformation. However commonsense and logical a proposed change may seem, there will be people within the organization for whom it does not cohere with their assumptions and therefore stirs up emotions leading to resistance.

Even after putting the line-stop authority into practice, the job of management is not done. As problems become visible, management must act with urgency to respond, engage in respectful problem solving dialogue, and implement countermeasures. This builds trust. In addition, because problems are detected earlier when they are still small and easier to correct, belief in the system increases, and this begins to embed the cultural assumption in the minds of the people. For organizations without an adequate level of stability or process support to start a stop-and-call system, the same effect of building trust in exposing and solving problems can be built through daily management and *kaizen* projects. The daily activity of listing and reviewing concerns, causes, and countermeasures (3C) on the team boards or the project activity of mapping processes and value streams to identify wastes and create action plans embeds the *kaizen* value that it is safe and good to make problems visible.

Respect for Individuals and Their Development

Many organizations today subscribe to the belief that talent follows a normal distribution across a population, the bell curve. Some have adopted variations of the 20-60-20 approach—to promote and challenge the top 20 percent, keep the middle 60 percent, and move out or somehow manage the bottom 20 percent performers. There are pitfalls to this approach. First, unless it is growing and hiring aggressively, an organization must define the top 80 percent by continuously finding the bottom 20 percent to move out of the organization. Unless the organization is growing, it is difficult to provide opportunities for people to grow and develop with it, and the search for the bottom 20 percent becomes focused not on motivation and development but on punishment. Second, it is questionable whether organizations that practice this approach have the interest or skill to measure not only the results achieved by the 20-60-20 groups but also the process they followed, and failing to do this, they could be rewarding luck and randomness, not the actual growth of talent.

Third, this approach creates a climate of fear that is incompatible with *kaizen* values. Toyota takes the responsibility to develop 100 percent of its talent rather than cut off the bottom 20 percent. This forces the organization to become better at developing people, better at finding the strengths of people, and better at rotating its people to roles in which they can succeed and contribute to the organization. Furthermore, Toyota's approach is to pull out the top performers and rotate them to other areas of the organization where they may be needed or where they can develop further by working toward new targets in new environments. This challenges the organization that was hosting these top performers to adapt to the loss of the top performers by promoting the best of the remaining 80 percent and by taking a hard look at the work and how it is being done—to look for *kaizen* opportunities so that the same work can be done just as well or even better without the star performers (Nemoto 1983).

Serving the Internal Customer as a Coach

The soccer team at Hiroshima Prefectural Aki South High School has been featured in the Japanese media several times because of the unique approach of its coach (Nippon Television 2013; Fuji Television 2013). For many years, corporeal punishment and physical violence from coaches toward athletes and students have been a troubling reality in Japanese schools. Not a few homepages of Japanese schools now have a link for students to report these incidents. In the schools' sports clubs, there is the traditional belief that tough practice builds character and that physical punishment helps students to learn. Some beg to differ. Coach Kimio Hata has led teams to high school national championships, and he has done so without corporeal punishment as one of his coaching techniques. "That creates order-takers" rather than people with initiative, according to Hata. His unique approach includes

▲ *Self-directed soccer practice.* Each student soccer team member decides what areas he needs to work on during practice. Coach Hata sometimes gives advice.

▲ *Three free days per week.* Only two days of practice are scheduled per week, with two other days of games, giving students time to do as they wish on the other days.

▲ *Self-directed soccer matches.* The team members decide who plays and the game strategy through discussion. The coach sometimes gives advice. "It requires great patience" not to give the answer, says Hata.

▲ *Discussing mistakes.* Hata does not tell the students what they did wrong. Instead, he asks them what they think they did wrong and what they think they should do.

▲ *Daily exchange of letters with all his students.* This coaching style may seem hands-off, but it is based on deep mutual trust and respect for the individual built up through the personal connection coach Hata makes by daily written two-way communication.

The coach does not score the goals. He does not even play on the field. The coach can give advice from the sidelines, but clearly, Hata has found that it is best to draw out the ability of his players to think for themselves in order to become better athletes, better competitors, and better young men. Perhaps we could call his approach "more with less" coaching. But looks can be deceiving, and we must recognize that Hata's approach to coaching requires more skill, care, and understanding of human psychology than the typical direct-and-drill approach. By removing blame for making mistakes, removing fear of physical punishment, giving the responsibility for learning back to the student, creating alignment with the team purpose to win games, teaching students important problem solving skills, and giving each individual his daily attention, Hata is putting his core beliefs of *kaizen* into practice in a unique and admirable way.

Did Mr. Nakao Just Say, "Two Mules"?

Organizations must place a high premium on understanding their processes in order to avoid shallow problem solving, jumping from solution to solution, ever fighting fires, without ever seeing the facts up close and finding the simple countermeasures to root causes. Worse, people will not develop these skills and the organization can fall into a cycle of superficial firefighting. As a result, people give up initiative, and it is left up to an authority or someone who claims to know what is best to decide how work gets done. This is frighteningly common worldwide. Even when people are aware of problem solving tools and skills, the process may not be understood correctly or from the customer's point of view. This brings about local optimization of processes, resulting in one process being fast or productive, whereas the end-to-end process cannot deliver according to customer needs.

Another common reason people struggle to put in place quality-at-source systems is that by stopping work to fix the problem, short-term losses are generated because the process is unable to deliver outputs. This is also an incorrect understanding of the process. As the same problems occur again and again over the long term, losses mount. These problems compound with other problems and cause overproduction, desynchronized processes requiring additional management effort and cost, and most deadly of all, the lack of learning. At other times, we may become very good at problem solving but unknowingly use inappropriate resources or methods. Understanding our processes requires an ability to both zoom out and see the big picture and zoom in on the fine details.

Mike Wroblewski recalls the surprising way that one of his *sensei*, Chihiro Nakao from the Shingijutsu Consulting Firm, tried to get people to grasp the lesson he was teaching. One day after coaching a *kaizen* event, Nakao was walking through the Hill-Rom factory with the senior management team. Mike recalls Nakao only saying, "No good!" and "Fail!" and "NG!" as he went from area to area, pointing out how Hill-Rom was not getting the idea. Rather than encourage, motivate, or recognize where his client was doing good work, the feedback to management was tough. What is more, Nakao walked quickly from area to area without explaining why it was "No good," leaving that up to senior management to reflect on and figure out for themselves, exhibiting a high-context communication style.

Wroblewski was prepared for Nakao's visit to the stretcher line, where Hill-Rom built recliner beds. He had been putting in place all the *kaizen* ideas he had been learning, such as attaching wheels to everything, using quick-connect power and air lines, using a circle-shaped line paced at a *takt* time of 75 minutes, and launching a home-made bicycle-chain train built with available materials. When Nakao saw the stretcher line, his comment was, "You passed, but just barely." It was the only area in the plant to receive a passing mark. However, it was Nakao's next question that surprised Wroblewski.

"Why did you spend so much money on the automation?"

Wroblewski was ready to explain that the conveyance system was built according to *kaizen* principles, using available materials, for less than $5,000. Nakao would hear none of it, instead asking, "Why did you think of automation first? Why not be more creative? It would have been better to have used two mules to pull the hospital beds around in a circle!"

Nakao was pleased with the result, but he was challenging the process. The stretcher line had passed, but just barely because Wroblewski had not understood the process deeply enough to gauge the need for automation correctly. For a line paced at a *takt* time of 75 minutes, what level and speed of automation was appropriate? How simple, slow, and cheap could the automation be? The unspoken value being taught was *use your creativity before using your capital.*

Acting with urgency, Wroblewski submitted a requisition for two mules to the purchasing department the same day, but it was rejected.

Urgency, Connecting, Consensus

Although urgency is much discussed by businesspeople, politicians, and academics in the context of change and getting things done, we believe that nearly everyone misses the point. What do we feel when we hear that we must act with urgency or that something is urgent? Anxious? Stressful? Fearful? Tense? Alert? Energized? Motivated? *Urgency* simply means calling for immediate attention. Yet, in common use, it has come to mean that there is a problem whose resolution is time-sensitive. There is an implied threat. Our response to threat may be to face it or to flee. It would seem that the *kaizen* value of urgency would conflict with that of security. But this is not so. *Urgency* simply means paying immediate attention. We cannot pay attention to everything, so we must pay attention to what matters, to what is most meaningful to our individual or shared purpose. In dire situations, this may be survival. If we believe in continuous improvement, what requires immediate attention is to make even small improvements daily. Urgency has an entirely overlooked positive dimension. When children are playing, it is an urgent matter. The play requires all their immediate attention. They are immersed; they are in flow. The same is true when adults are working on hobbies or on interesting projects or when they are with ones they love. These are urgent matters that we look forward to, make time for, and that gives us energy. Another way of thinking of the *kaizen* core belief of urgency is any strong source of motivation and energy.

The topic of connecting processes has perhaps been written about and studied more than any other aspect of the Toyota Production System. Indeed, the early understanding of the entire system was summed up in the words *just-in-time*, and for many years this was misunderstood to be the total system. Today we know better. Even with the expanded understanding of the entire structure of the Toyota Production System and the material and flow interconnections made visual through value-stream mapping, our understanding of what it means to connect people, processes, information, and material movement is inadequate. There are three main aspects to connection. The first is the simplest and involves the reduction of inventories and total cost by working in a flow. Liker (2003), Womack and

Jones (2003), and others have explored this at length. The second aspect is that removal of barriers to connecting processes, be they physical, policy, or transaction costs related to economies of scale, results in building systems that expose these underlying problems. Often these economies of scale are deceiving, merely justifications for now connecting processes rather than true economies. The third aspect is to connect people and the work they do through natural work teams or cross-functional teams or simply by building stronger human relations in general across the organization. This also enables people to expose and share problems and concerns quickly using tools, including but not limited to standardized work, hourly plan versus actual boards, project review meetings, and strategy deployment reviews. The Welsh proverb, "He who would be a leader must be a bridge," speaks to the role of the leader in facilitating connections.

As part of this standardization and codification of how improvement was done at Toyota, the eight steps of the plan-do-check-act (PDCA) cycle were detailed in the Toyota Business Practice (TBP). Even these eight steps are high-level guidelines, and we can examine each one more deeply to see that in fact there are major steps and key points to check at each step of the way. Great emphasis placed on creating well-defined problems is evident in that five of the eight steps address the plan phase (Table 3.3).

This same basic process is followed for all variations of the PDCA cycle, be it daily *kaizen*, project *kaizen*, or long-cycle support *kaizen*. The emphasis on slow and thorough planning with clear problem statements, clear targets,

Table 3.3 Consensus Building Within the PDCA Cycle

PDCA Stage	8-Step Practical Problem-Solving Model (TBP)
Plan	1. Clarify the problem
	2. Break down the problem
	3. Set the target to achieve
	4. Analyze root causes
	5. Develop countermeasures
Do	6. See countermeasures through
Check	7. Check process and results
Act/Adapt	8. Standardize successes, learn from failures and identify gaps for next plan

adequate root-cause analysis, and consideration of a variety of counter-measures gives both time and process to build consensus among team members and stakeholders in the proposed change. The result is rapid and sure execution of the plan. We will explore the implications this has for the quality of decision-making and strategy deployment in Chapters 4 and 5.

Sharing Goodness Beyond the Organizations

Many large corporations today encourage their employees to volunteer in charity organizations and donate funds and time. The *kaizen* value of sharing builds on this—to share the knowledge, skill, and passion of the people in the organization to help others in the community improve. This is the proverbial "If you give a person a fish, he or she will eat for a day, but if you teach the person how to fish, he or she will eat for a lifetime." Companies such as Intel, Toyota, Boeing, and others have sent their employees to teach food banks how to use *kaizen* to streamline their processes and do more with less. The "Corporate Citizenship" section of the Boeing website reported the following (Boeing 2009): "Zoos, hospitals, schools, engineering projects in remote areas—anywhere people are trying to do a good job—that's where Boeing employees bring value-added skills to make the world a better place," said Anne Roosevelt, vice president for global corporate citizenship. Indeed, Boeing has been instrumental in launching continuous improvements in many organizations local to its many manufacturing and engineering sites, sending its lean manufacturing experts free of charge to help make improvements and teach *kaizen* in support of continuous-improvement efforts to local government functions at King County and the State of Washington. The customer-supplier continuous-improvement efforts are not always easy to start because the commercial relationship creates potential conflicts of interest, brings up issues of trust, and spotlights differences in organizational culture. Yet it is essential for a company's long-term prosperity to extend continuous improvement across all its customers and suppliers, connecting everyone. Sharing good practices within the community also presents a good opportunity to develop in-house skill in bridging such interorganizational gaps.

Beliefs, Decisions, and Our Destiny

The level of integrity and commitment to these core beliefs will differ for each person. How we put them into action will be totally different depending on whether we are doing it because "engagement" is a metric tied to individual performance plans or whether it is something we passionately believe, ideally from experience. The actions of leaders speak louder than words. When shaping a culture, the desired core beliefs and behaviors need to be defined and spoken explicitly. This begins with humility, alignment, and a safe environment in which to point out, "Hey leader, your behaviors aren't reflecting our core beliefs."

It is up to the leaders of an organization to make decisions about future directions. This is an awesome responsibility. Too many times the way decisions are made is opaque, the process is not defined, and the outputs are not clearly linked to the inputs. *Kaizen* transformation involves leveraging core beliefs of *kaizen* and the behaviors they generate to become better at making decisions.

CHAPTER 4

The Meta-Decision
of *Kaizen* Culture

In the struggle for survival, the fittest win out at the expense of
their rivals because they succeed in adapting themselves best to
their environment.

—CHARLES DARWIN

One of the greatest challenges faced by organizations attempting to
transform themselves by systematically pursuing excellence is the deeply
embedded set of assumptions and ways in which we determine truth. For
individuals, determining truth can be an important part of finding their
purpose or way in life. Likewise for organizations or groups of individuals
who have come together for a shared set of objectives, we can say that the
need to determine truth is the need to find the right course of action both
in day-to-day and long-term strategic matters.

Determining truth for organizations is not cosmology or a spiritual
quest; it a question of making good decisions, of finding the way forward. For
organizations, the objective of determining truth is to decide on a course of
action in response to environmental conditions and in pursuit of long-term
ideals. Futurist Alvin Toffler (1984) wrote that to avoid what he called "future
shock," we must become "infinitely more adaptable and capable than ever
before. We must search out totally new ways to anchor ourselves, for all the
old roots—religion, nation, community, family, or profession—are now
shaking under the hurricane impact of the accelerative thrust. It is no longer
resources that limit decisions, it is the knowledge that makes the resources."
Organizational cultures, as the sum total of their assumptions, values,
behaviors, and artifacts, act as a filter to the information presented. The

Figure 4.1 Organizational culture as a decision-making filter.

filtering of reality or information about reality through the organization's culture is what determines the quality of decisions (Fig. 4.1).

Ironically, how we make decisions hardly seems to matter to the senior leaders of many organizations. We are measured on results, and as long as the results are good, we don't question how we arrived at the decisions, or even whether there is correlation between management decisions and results. A core belief of *kaizen* is the importance of checking both *process and results*. Within a *kaizen* culture, changes are conducted as experiments, as hypotheses to be tested, rather than positions or opinions to be defended. We are as interested in the process of making and implementing our decisions as we are in the result that they yield. When the results are good, it is important to know whether it was due to luck or due to following a good process. When the results are not as we hoped, it is an opportunity to learn through trial and error, to become better at finding truth. By reflecting on both the process and the results of our decisions and applying *kaizen* to future decisions, we become better at making decisions.

Schein (2004) described six ways that organizations decide what is true and find their course of action in his "Criteria for Determining Truth." We have built onto this an additional "adaptive" level to these criteria (Table 4.1).

Table 4.1 Criteria for Determining Truth

Approach	Description	What We Might Hear
Pure dogma	Tradition, religion	"We have always done things this way."
Revealed dogma	Wisdom based on authority	"The boss (the most senior or most experienced person) says this is the way."
Rational-legal process	Socially-determined truth, not absolute truth	"The majority (the review board, committee or responsible department) has spoken."
Conflict survival	Truth is what survives the process of conflict or debate	"Let the best argument win!"
Pragmatic	Truth is what works	"Let's go with whatever works."
Scientific	Truth established through the scientific method	"If it doesn't agree with experiment, it's wrong."
Adaptive	Recognizing that all six are necessary and valid, being self-aware in selecting the appropriate approach for the situation, testing and validating the outcomes	"Let's understand the nature of the situation and choose the approach. If we are wrong, we will adjust."

(Adapted from Schein 2004.)

Organizational Culture as a Decision-Making Filter

The six criteria described by Schein can be viewed as a progression from decisions made from dogma to decisions emerging more scientifically. Indeed, in today's industrialized world, science has brought great progress in the lives of people and in how businesses are run. However, this does not mean that one day all decisions will be made devoid of tradition, respect for authority, debate and social consensus, or what is pragmatic for the occasion. We are not advocating scientism. We propose that the highest level of decision-making within an organization must be *adaptive* and encompass the ability to select from all six approaches as fits the situation.

Decision-Making Based on Tradition

Whenever we hear, "because we've always done things this way," we know that tradition is a strong criterion for decision-making. Although pure dogma is seldom the main or only decision-making approach within successful modern organizations, it is important to value and respect tradition for several reasons. First, science is only just beginning to speak on the subject of biology and morality. Studies have shown that as early as 15 months, babies demonstrate a sense of fairness and altruism (Schmidt and Sommerville 2011). Scientific studies may someday prove that our religious or traditional morals also have scientific underpinnings. Until that day, we have to trust and respect our feelings about what is right and wrong and weigh them against what the cold facts tell us to do when deciding a course of action.

Second, tradition and even ritual can bind people in positive ways to the purpose of an organization, giving it meaning and cohesion. In a *kaizen* culture, we should respect traditions held in value by people, but at the same time we should test them pragmatically and scientifically to see whether they are still fit for the times or require adaptation. Changing traditions can be difficult, but history shows that when we do not embrace and take control of change, change takes control of us.

Third, organizations are human endeavors and not products of pure science. Some organizations exist mainly for social purposes or to further causes and do good in the world. It may not always be possible to rationally or scientifically show why certain business decisions are good ideas. We should do our best to find fact-based underpinnings of important decisions and to weigh the results after following our actions, but it is neither always possible nor desirable to disconnect how we decide from how we feel. The *kaizen* core belief of respect for individuals and groups requires us to understand traditions and how they have helped to keep people aligned to common purpose, created a feeling of security and connection to one another.

Decision-Making Based on Authority

When one person's voice consistently carries more weight than that of the group, or when people expect this voice to have disproportionate influence, this is a sign of authority-based decision-making. In its simplest form, authority-based cultures are not compatible with adaptive cultures, which require a greater degree of engagement from all levels in gathering knowledge and making improvements based on many fast experiments. Revealed dogma or the accumulated wisdom of an authority figure should never be completely neglected in favor of science, which some would say is just a different authority figure or, at worst, a dogma. At different points in the development of scientific knowledge, we may not have explanations for wisdom such as "An apple a day keeps the doctor away" because there are far too many such truths to study and far too few scientists to perform the studies. In the absence of scientific evidence, it is prudent to listen to your grandmother or "do what worked in the past" and keep eating apples. Evidence should be respected in decision-making, but we must remember that the absence of evidence is not evidence of absence and listen to the voice of experience when this is the best available proof.

It is also important to remember that there is a time and place to follow the leadership of authority figures, even for "command and control" style leadership. Organizations faced with an urgent need to embark on a major change effort can rarely do so successfully without a strong leader taking command with the message of "We must change!" or "We must adapt!" and eventually, "We must do so scientifically!" These are not conclusions to which groups can reliably arrive through consensus-building discussion, and even when this is possible, there is not always time to do so when survival is at stake. Leaders must be able to persuade people with urgency through a combination of logic and rational argument, credibility and character, and emotional connection with the listener. It is not reasonable to demand that decisions be made purely rationally because people will not follow decisions that are rational yet too difficult to integrate emotionally, a topic of more discussion in Chapter 7.

Decision-Making by "Majority Rules"

An organization relying heavily on a rational-legal approach to decision-making often enjoys a high level of consensus and engagement when it comes to execution of the decisions because of the fact that they are made by or close to those responsible for execution. This can be seen in the "plan slowly and thoroughly and act quickly" principle of the Toyota Way. Clear lines of responsibility and agreement on the process of consensus building are prerequisites for this approach, as well as the environment where it is safe to make these decisions and their consequences visible to all. The *hoshin kanri* process, which will be described further in Chapter 5, and related visual management methods are artifacts of this style of decision-making. But there are risks when the rational-legal approach becomes the strongest or most common style for seeking truth and making decisions about the way forward. In organizations lacking a common approach to planning or problem solving, the rational-legal approach relies on people in positions of responsibility being able to make good decisions. Even when people in positions of responsibility are competent, personal styles of decision-making vary heavily from dogmatic to scientific. Sometimes there are absolute truths such as "What is the most effective known ways to sell product X in market niche Y?"

The rational-legal approach may not arrive at this truth successfully because determining the way forward is a matter of following the rules left to the responsible party rather than seeking objective truth by review of available data and anthropological surveys of customer behavior.

This inconsistency in how people make judgments can cause friction and misalignment and is countered within a *kaizen* culture by establishing an agreed basic framework for problem solving and decision-making such as the plan-do-check-act (PDCA) cycle. The most typical pitfall we see with the rational-legal approach is that an organization becomes overly bureaucratic, requiring many reviews, signatures, and approvals even for decisions that are quick, commonsense issues. In these types of organizations, such rules and processes have become shields for people against blame when mistakes are made. As a result, the worst case can be where people disengage from thought itself, allowing complex or excessive rules to promote behaviors that are less than fully intelligent. This is what Hall (1976) referred to as "extension transference" or "*kaizen* for the sake of *kaizen*" and not for the sake of improving what is truly important.

The *kaizen* culture countermeasure to this big risk of the bureaucracy or framework becoming the end in itself rather than serving the purpose is not to say, "The fewer rules, the better," but to insist that rules and standards be pervasive, well documented, and constantly improved. This is counterintuitive and even displeasing to many who expect more freedom from rules and policies within a more "lean" work environment. What is often missed is the speed, frequency, and levels at which rules are updated and improved. Within a *kaizen* organization, rules and standards are always seen as provisional and subject to revision and improvement as soon as the environment changes or as soon as better methods are found. There is a famous story about Taiichi Ohno, architect of the Toyota Product System, observing on one of his *gemba* walks that a standardized work document had not been updated in over a month and scolding the person responsible that he was a salary thief, not doing his job of continually looking for better ways (Ohno 2012).

Decision-Making Based on "Trial by Fire"

The conflict-survival approach to determining truth has many of the same risks and benefits as the rational-legal approach. In the conflict-survival approach, truth is what survives the process of debate. Although *conflict* may give a negative impression, a dialogue-based approach to deciding on truth and best courses of action within an organization has many benefits, such as challenging embedded ideas, traditions, or the status quo; openness to considering opposing ideas; and inclusiveness towards multiple stakeholders in the decision-making process rather than just a bureaucracy, committee, or department. The organization with strong traditions and authority figures is cautioned to practice the rational-legal approach by setting rules for debate in order to avoid conflict survival turning into "decision by authority" or real conflict. The *kaizen* core belief in understanding who we serve is practiced by viewing the issue or the work we do from the point of view of the customer. This requires organizations to think across functional boundaries and find the most appropriate end-to-end answer rather than the answer that makes the most sense from a functional (e.g., sales, quality, or logistics) point of view. In an environment where it is safe and secure to make issues visible with the actual situation, connecting people in teams to directly observe and map processes can be an extremely powerful way to make good decisions while building alignment and consensus.

Pragmatic Decision-Making

The pragmatic approach involves simply trying different ways and going with what works. The obvious advantage is that the decisions made or options selected are the ones that provided results. The not-so-obvious disadvantage is that because the approach is not yet rigorous and scientific, selection of the way that works may be due entirely to luck. The *kaizen* core belief in the merit of understanding your processes can be violated when a purely pragmatic approach to decision-making is followed, and success may not be repeatable. Organizations can spend tremendous amounts of energy trying different approaches with varying degrees of success. Individuals who have been lucky a few times may be rewarded, becoming an authority who then makes decisions via the "revealed dogma" approach without any repeatable process for obtaining truth. Too many organizations reward only results and not process, entrusting long-term success of an organization to senior leaders who may have survived on luck with the pragmatic approach. Still, learning by doing is a powerful way of making decisions, especially when it is done with urgency, as a team, and with a spirit of humility to learn.

Scientific Decision-Making

The scientific approach establishes truth by collecting and weighing evidence through the open-minded inquiry of the scientific method. This takes the form of experiments designed to rule out inferior theories. Unlike dogma, science recognizes that knowledge and understanding are incomplete, that the process of scientific inquiry will probably never be finished. The scientific outlook requires a certain level of comfort with ambiguity and uncertainty, even in the face of incontrovertible evidence. Future evidence may be found to revolutionize our understanding, and indeed, in the history of science this has happened many times (Table 4.2).

Table 4.2 Changing Scientific Paradigms Through History

Past Scientific "Truth"	Current Scientific Truth or Dogma	Future Scientific Truth?
Earth is the center of the universe	Earth is not the center of the universe	There are many universes...?
Animals are not intelligent	Animals have intelligence	Animal-human translation...?
Matter is made of atoms	Matter is made of particles	Dark matter is...?
The brain does not change	Neuroplasticity	The brain is capable of...?
Heart is seat of consciousness	Brain is seat of consciousness	Consciousness may be an emergent phenomenon of complexity...?

In this regard, it is quite different from belief based on tradition or experience that serves to give us certainty and comfort, even though our beliefs may be wrong. The major difference between a *kaizen* culture and a more traditional organizational culture is in the insistence on managing by facts. The scientific approach is to look for evidence, and evidence is extracted from facts. Facts are not always convenient, nor do they always agree with our beliefs. We are not wholly mistaken to question facts, but we must *trust but verify* by going to the source and seeing for ourselves and then testing our theories through well-designed experiments.

Science, when done properly, is self-correcting and nondogmatic. Nobel Prize–winning physicist Richard Feynman said, "It doesn't matter how beautiful your theory is; it doesn't matter how smart you are. If it doesn't agree with experiment, it's wrong." In reality, human failings cause the practice of science to be far from scientific all the time. This is no exception within organizational cultures because people can disguise authority or dogma-based decisions in scientific language, data, and analysis with poorly designed or even manipulated experiments in order to win arguments and influence. Truly, the transformation to a *kaizen* culture requires that we use the scientific method not as a tool but as a way of thinking. This can be a deep personal change requiring strong leadership commitment, opportunities to learn from mentors, and environments within which to experiment and fail safely.

Statistician, sabermetrician, psephologist, and writer Nate Silver (2012) argues that we need to improve our ability to distinguish between "what we know and what we think we know" because even highly trained scientists, statisticians and authority figures are led astray by this gap. Much like the *kaizen* process, Silver advocates making and testing hypotheses frequently in order to address the cognitive gaps of overreliance on past data, placing too much weight on the most recent data points, and failing to take into account new facts when they do not fit our beliefs. He writes, "The more often you are willing to test your ideas, the sooner you can begin to avoid these problems and learn from your mistakes."

Many people misunderstand or may be intimidated by science, and we should be aware of its limitations. Although science is the original and ideal of open-minded inquiry, scientism is a close-minded and reductionist attempt to deny other avenues of arriving at knowledge and truth. Just as author, scholar, and mathematical investor Nassim Taleb (Economist online 2010) describes himself as "someone who uses probability to show where you cannot rely on probability," we must use the scientific method to discover the limit of this approach to decision-making in the world of business, nonprofits, or other enterprises where we are faced with inadequate information, imperfect people, and not enough time to run well-designed experiments. We must allow ourselves the choice to adapt and select the next best option, whether it is consensus, a leader's direction, or pragmatism, whatever seems to work. In fact, Taleb (2012) argues that decision-making methods that are entirely heuristic or experience-based and nonprobabilistic are essential to building what he calls "antifragile systems." For example, traditions that have survived the test of time can be very useful in decision-making; as Taleb says, "Time is the best statistician."

Adaptive Decision-Making

Therefore, the final and overarching criterion for determining truth, as an expansion to Schein's model, is what we call the *adaptive approach*. The adaptive approach recognizes that although evidence-based decisions are desirable above all whenever possible, each of the six criteria is a necessary and valid approach both for individuals and for organizations. The adaptive approach views the six criteria not as a linear progression ending in the scientific method but as ways of making decisions based on situations and environments. This requires that we become more self-aware in selecting the appropriate approach rather than simply falling back on tradition, beliefs, the voice of authority, or the committee. While these are all valid, none of them is the answer by itself or for every situation. The *kaizen* culture is above all else practical, realizing that it is not possible to always rely on scientific evidence when making day-to-day decisions. However, in a *kaizen* culture, we must always make time to return to the scene of the decision and insist on facts to validate or disprove that it was the right decision in order to learn and improve how we decide in the future.

Liker (2003) identified five elements within the decision-making process that are part of the Toyota Way. This approach is, by our definition, adaptive, selecting a combination of decision-making styles:

1. Find out what is really going on, including *genchi genbutsu*.
2. Understand underlying causes that explain surface appearances—asking "Why?" five times.

3. Broadly consider alternative solutions, and develop a detailed rationale for the preferred solution.
4. Build consensus within the team, including Toyota employees and outside partners.
5. Use very efficient communication vehicles in styles 1 through 4, preferably on one side of one sheet of paper.

We see the scientific approach in styles 1 and 2, the rational-legal approach in style 3, conflict survival in style 4, as well as possibly also revealed dogma with the inclusion of internal or external subject matter experts. Style 5, which is not decision-making per se but communication, has a pragmatic element to it. The *kaizen* core beliefs in evidence are building alignment, understanding the process by *genchi genbutsu*, respect for individuals and stakeholders in the decision-making process, building consensus, connecting internal and external people into the team, and providing security by reducing psychological stress through efficient communication using the artifact of the A3 report.

In summary, the transformation toward an adaptive, continuous-improvement-minded organizational culture means engaging in decision-making processes that strengthen *kaizen* core beliefs. These are basic habits that are very difficult to change but that have an enormous impact on success or failure for individuals and organizations. At the very least, we must be open to change, stepping out of our paradigms that tradition and authority provide all the answers. We must engage the collective experience and wisdom of people in the organization and, further, take a pragmatic and scientific approach to validating our most important decisions and their outcomes. Finally, we must take an honest look at how we decide whether we are achieving the results we desire in our organization and the lives of the people within it, and have the courage to adapt to a better way. We must begin with humility.

From "My Way or the Highway" to the Kaizen Way at HON Industries

Established in 1944 and headquartered in Muskatine, Iowa, HON Industries designs and manufactures office furniture, including chairs, filing cabinets, workstations, tables, desks, and educational furniture. Experiencing strong growth and financial performance, the journey to transformation at HON began in 1992 when customers began demanding greater speed, flexibility, and responsiveness. Study missions to Japan, leadership education, and shop-floor improvement projects led to the development of the company's own lean manufacturing, the HON Production System. Through this system, HON came to practice the PDCA cycle at all levels of the organization—through policy deployment at the strategy planning and monthly, quarterly, and annual performance-management levels; through *kaizen* events at the project level; and through daily team status-check and action cycles. With periodic benchmarking of other industry leaders and development programs to build internal expertise in *kaizen* and lean manufacturing systems at all levels, the HON Production System remains a pillar of the company's competitive strategy.

Steve Burkhalter, who spent 14 years in front-line management at Toyota Motor Manufacturing Kentucky, left in the year 2000 so that he could share what he had learned about *kaizen* and the Toyota Production System with other companies. In 2005, he joined HON Industries and was put in charge of transforming some newly acquired production sites. His first step was to spend time in the plants to understand the customers, the people, the processes, and how they were managed. He used a detailed assessment that scored against criteria of leadership, empowerment, vision, innovations in services and products, partnering with suppliers and customers, environmental practices, manufacturing processes, customer satisfaction, safety, quality, delivery, cost, productivity, and profitability. "The factory in Paoli, Indiana, had a very old school authority-based management style, 'My way or the highway,'" recalls Burkhalter.

Aligning with corporate strategy and goals, setting a foundation of security with a "no layoff policy" as a result of any improvement activity, and gaining a solid understanding of the process, Burhkhalter and his team built a plan of action down to the individual team member level. "*Kaizen* actions were linked to policy deployment at the plant level, and at the plant level to the group level. This opened up empowerment and ideas from people on how to meet these breakthrough targets," says Burkhalter. Daily *kaizen* and follow-up on visual performance metrics were led by lean engineers who had zone responsibility to work with supervisors and managers to improve the performance in their areas. Between four and six *kaizen* events were held each month involving cross-functional teams. Recalling the level of top-management commitment and involvement, Burkhalter says, "Stan Askren, the CEO, would participate in the *kaizen* events, especially the plant manager *kaizen* events, which focused on business processes such as order entry. These were as much for learning by the plant managers as they were to redesign the processes targeted by the *kaizen* team." Over a period of four years, this well-supported, well-executed *kaizen* transformation made sustainable high-double-digit improvements across all performance metrics.

Why *Kaizen* Transformations Succeed

As a process, *kaizen* transformation has been tested and refined over decades. Although there is no single universal approach, there is general agreement among practitioners on factors that cause transformations to fail. Kotter (1996) illustrated why organizational transformation efforts fail and described the eight steps that work as countermeasures to these failure factors.

1. Establish a sense of urgency.
2. Form a powerful guiding coalition.
3. Create a vision.
4. Communicate the vision.
5. Empower others to act on the vision.
6. Plan for and create short-term wins.
7. Consolidate improvements and produce still more change.
8. Institutionalize new approaches.

This is a combination of planning and preparation, execution of the changes, and following through. Steps 1 through 4 describe communicating the purpose and importance of change. Steps 5 through 7 are about the changes themselves, about creating wins and building credibility for the change effort. Step 8 is about embedding the values, assumptions, and behaviors deeply in the culture so that the transformation will be successful and lasting. *Kaizen* transformation succeeds because in a very practical way the *kaizen* approach takes into account many, if not all, of the critical change-management steps required for sustainable positive change.

Kaizen changes culture because artifacts in the form of daily maintenance-and-improvement routines reinforce behaviors, and behaviors develop into assumptions; then these assumptions generate behaviors that create new artifacts, and the cycle goes on. At the same time, the entire way of working is established not as dogma or belief but as a hypothesis that is tested and refined daily through experimentation. Does making problems visible create more trusting positive dialogues between people? Does this result in faster and more permanent problem resolution? If not, is the process of exposing problems flawed, or is our execution of that process flawed? And so forth as the PDCA cycle challenges assumptions and the process-to-result relationship of each of the root-level *kaizen* values. This empowers and challenges people to continually renew the system, which, in turn, creates a fun and motivating workplace.

Kotter (1996) wrote: "In total, all of the practices I've been describing here will help an organization adapt to a rapidly changing environment. Creating those practices so [that] they stick is an exercise in creating adaptive corporate cultures." Although Kotter described the eight steps and practices, it was left to the reader to develop the "how," or practical methods, for taking each of these steps in a meaningful way. The examples and illustrations that follow in this book will reveal how organizations are sustaining *kaizen* transformations through concrete, practical actions to make better decisions and become better at decision-making.

Broad-Based Skill Building at Chrysler Corporation

Since the 1980s, we have become accustomed to hearing about *kaizen* and world-class manufacturing practices being transferred from Japanese companies to the West. In the past decade, this transfer of knowledge has increasingly been between and among Western organizations. This is both natural and desirable, evidence that *kaizen* itself is adapting to various cultures and being transmitted worldwide. When the financial crisis of 2008 struck, it became clear that the American automotive industry was particularly unprepared to adapt, in terms of both the agility of its supply chain and its organizational cultures. Chrysler Corporation was one such company that was forced to file for bankruptcy.

Chrysler rapidly emerged from bankruptcy and has started down a path of cultural transformation as a result of knowledge transfer in world-class manufacturing from an unexpected place: Italy. When Chrysler adopted Fiat's lean-based World Class Manufacturing System in 2009, this represented a fundamental shift in how the company's plants operated. As

Scott Garberding, senior vice president of manufacturing and world-class manufacturing at Chrysler explains, making this happen was all about developing people.

> The World Class Manufacturing (WCM) System, which we got from Fiat, has changed every facet of the way we build cars, making big differences in the way work gets done, who does the work, and the way the work is led. WCM was a cornerstone of Sergio Marchionne's turnaround strategy when he assumed the CEO role in 2004, and the deployment of WCM was one of the first efforts initiated between Chrysler and Fiat. I've been with Chrysler for about 20 years, mostly in manufacturing, and from my observation, we've made more change in the last 3 years than we did in the prior 10 or more.
>
> Briefings about WCM began even before we emerged from bankruptcy, and when employees came back to work in June 2009, there were signs of WCM in every one of our plants. However, when I returned to manufacturing in October—I had headed up purchasing through the bankruptcy—I saw that the changes were fairly superficial compared to what I had seen in the Fiat plants. There were aspects of our work that we needed to start tearing down immediately in order to change as we needed to change.
>
> So I began some rather emotional exercises with my senior leadership team that we called "Destroy to Build." Essentially, I was asking people to move outside their comfort zone.
>
> We began by immediately asking the groups to put together plans to comply with WCM audit criteria. This was extremely aggressive—probably too aggressive considering the learning curve our people were facing—but it got the plants certainly motivated and thinking about moving forward. I went to all of the audits—every one of them—and stayed for the entire sessions in order to gain a thorough understanding of the methodology, and to demonstrate that this was very important to us and the way we were going to run the business.
>
> I also assigned—and this generated some angst—one of WCM's 10 technical pillars to each member of my staff, including the heads of HR and finance, and to myself. I tried to assign pillars

that were somewhat disassociated with people's primary function. Each staff member was responsible for teaching his or her pillar to the other staff members, and so they really had to learn how their pillars worked in the company.

We then went out to the plants, working systematically in contained areas with the key pillars. The idea we really pushed there was to learn. For me, one of the lessons was that taking an implementation idea from a fairly advanced plant and applying it to one just starting out with WCM didn't really work—the solution didn't fit exactly, and the knowledge wasn't there to apply and adjust it.

What we came around to was letting the plants iterate. We just encouraged them to keep moving. Some projects went through perhaps five generations of solution in a year's time, but when that was done, you had a core of people with deep skills that understood exactly what they had done and why they had done it, and who were then able to spread the methodology more widely in the plant.

All this has required a different kind of thinking than we were used to at Chrysler. At first, the team wanted to chase audit points, and some of the traditional high performers had a hard time giving up their own "secret sauce" and accepting the methodology. But over time, the audit scores showed that those who had latched onto the methodology were improving faster than those that didn't, and it became very clear to everybody that WCM really works.

Our leaders have had to meet tremendously challenging targets, but we also reward them for sharing their knowledge, and sometimes their best people, with other plants.

Essentially, there has to be the right balance between competition among the plants, which will always be there and should be, and maintaining a behavior set that supports everyone improving. Our management appraisal process now reinforces this by emphasizing 16 leadership behaviors that Sergio Marchionne previously rolled out at Fiat.

The overriding effect is that we have built skills in a much broader base of people than what we have ever attempted before. We have hourly team leaders who we now use as consultants to travel among the plants. They can talk about real experience and real barriers and how they got through them. Having that sort of larger army to go and attack problems has been fantastic.

There are aspects of this that most of us miss, particularly in North America. For example, I had a discussion early on with Professor Hajime Yamashina, who had helped Fiat build the WCM methodology, about suggestions from employees. Our thinking had been that you collect a huge number of suggestions, winnow them down to the critical few, and then implement those and get a big benefit. He corrected me, saying the real reason you want to drive a high volume of suggestions is that otherwise, you never ask your hourly workforce to put together a logical description of a problem or use a logical process to solve a problem. So the real goal is to develop your people and help them think differently about the processes they work in.

From my personal observation, a methodology like WCM is a big undertaking, and it is very different from the way most companies work. It has to be embraced, supported, and understood by the CEO because without that, there's just not the constancy of purpose, the determination in place to really make the shift.

As well, it is not something that can be learned from a distance. It requires direct participation on the shop floor from executives at all levels. Leaders have to understand the barriers that employees in the plants face because sometimes what we think are the barriers and what they really are, are two different things.

I would encourage leaders following a similar path to really focus on people development. I am amazed regularly at the depth of analysis that our teams do and the clever solutions that they are able to come up with. Sometimes these folks are engineers, and sometimes they are people who just have worked in the process for a lot of years. So you arm them with some effective problem-solving and improvement tools, and it is amazing what they go and do.

The WCM transformation at Chrysler did not begin as a culture-change initiative; it was launched to deliver the performance improvements necessary to rebuild the company. However, it is clear that the shifts in beliefs, assumptions, and values about work occurred in the process. In the

Chrysler story, several *kaizen* core beliefs in evidence include creation of an environment that tolerates failure and encourages learning by iteration, leaders who act as teachers and coaches, broad-based development of people and their problem solving skills, and recognition of the importance of process and not only results—as seen in the insight about quantity versus quality of improvement suggestions. The transformation at Chrysler is ongoing, underpinned by strong leadership and vision from the CEO, a sense of urgency, clear communication of the expected leadership behaviors, examples of success to inspire others, and recognition by leaders such as Scott Garberding that it is an emotional journey, not merely a change in technical or operational processes and methods.

The Strategic Decision to Create a *Kaizen* Culture

Organizational cultures are decision-making filters for not only day-to-day chores but also strategic choices of what products or services to offer at what prices, where to do business, who to employ, and how to manage their operations. In a free-market-based society, the decisions that organizations make are constrained by other decision makers—customers, investors, or shareholders, and the talent pool within the society. An organization must make decisions that are appealing to these three groups in order to remain competitive. When organizations in a free market fail to consistently offer good products to the market, good returns to shareholders, and attractive workplaces, they inevitably fail to excel.

However, even providing great products and services to the market, outperforming the stock market, and being a workplace of choice are not sufficient for sustained success. Long-term strategy requires attention to making the enterprise not only high performing but also sustainable, capable of surviving through adaptation. This is more than simply making a decision to become adaptive—to adopt a *kaizen* culture. This is a meta-decision—a decision about how we will make decisions in the future. In Chapter 5 we will show how building *kaizen* core beliefs into strategic decision-making and strategy-deployment processes and learning by doing are key factors in creating alignment and engagement in the transformation.

CHAPTER 5

Kaizen as Strategy in Practice

Not seeing a tsunami or an economic event coming is excusable; building something fragile to them is not.

—NASSIM NICHOLAS TALEB

Peter Drucker studied and advised leading organizations for more than half a century. He stressed that serving the customer well was the essential condition for prosperity, coined the term *knowledge worker*, taught respect for the worker, invented management by objectives (MBO), and had a great influence on management thinking in Japan and the West. His writing and insights contributed immensely to the field of management, which one could argue he invented. When he wrote, "Culture eats strategy for breakfast," he put into five words a powerful idea (Fig. 5.1). Our habits, routines, decision-making patterns, and unspoken group norms would make quick work of our best-laid plans that came from the hardest work of our best minds. This is not to say that strategy is unimportant or inferior to culture. Without strategy, even an organization with the best culture is simply adrift, reacting and responding to changes.

Furthermore, the ability to develop long-term plans, execute small experiments quickly, review the results of these experiments, and reflect on what was learned in those experiments and strengthen the strategies is an outcome of a *kaizen* culture. Culture only eats strategy for breakfast when culture is a monster that, untrained and unrestrained, has grown too strong without reflection and self-improvement. A *kaizen* culture is better-behaved and has better things to do than break its fast on our strategies. A *kaizen* culture tinkers with and continuously improves the strategy through daily

Figure 5.1 Culture eats strategy for breakfast.

action and review cycles. *Kaizen* seeks to understand the target, grasp the current situation, evaluate the gap between target and actual, take immediate concrete action to correct the gap, and learn from this process before repeating the cycle. This is repeated at cycles that may be within the day, week, month, or year depending on the scope. When the *kaizen* process is applied to strategy, the plan-do-check-act (PDCA) cycle is turned at these regular intervals so that plans can bend but not break.

Strategy Is a System of Expedients

Field Marshal Helmuth von Moltke the Elder was chief of staff for the Prussian army and one of the greatest strategists of the late nineteenth century. He was suspicious of rigid, inflexible, and totalizing grand strategies and theories, instead advocating strategy as a series of options that could be adapted to fit the situation. He coined the often-repeated phrase, "No plan survives first contact with the enemy." It would be wrong to think that Moltke thought that plans were of no use; his planning for war was very detailed and took into account a great number of variables (Bucholz 2001). He called strategy a "system of expedients" and said, "It is the translation of knowledge to practical life, the improvement of the original leading thought in accordance with continually changing situations."

In the context of business strategy, we should understand *expedient* to mean actions that are appropriate to the particular circumstances and that will deliver the results required, as in setting a general direction and adapting based on the actual situation after launching the plan. This would seem highly practical and logical, yet how many of us set, or have set for us, aggressive goals that turn out to be unreasonable on understanding the real situation, yet we are organizationally or emotionally unprepared to make expedient adjustments? Excellent plans that cannot be executed are a hallmark of nonadaptive cultures. Moltke believed that military strategy must be understood as a system of available choices because it was only the beginning of a military operation that could be planned. For him, the main role of military leadership was to make thorough preparation for all possible outcomes. We can summarize "strategy is a system of expedients" applied in business, and in particular for *kaizen* as strategy in action, in three points as:

1. Prepare plans thoroughly.
2. Put the plan in action, and gather information at the source.
3. Respond to reality, and adapt the plan.

Army general and thirty-fourth U.S. President Dwight Eisenhower said, "Plans are useless; planning is everything." Strategy as a system of expedients is not an action plan but a dynamic thing, a hypothesis to be tested on the field of battle, a basis for learning. It is being ready to adapt.

Hoshin Kanri: The PDCA Cycle Adapted to Strategy

Hoshin kanri (also called *hoshin planning*, policy deployment, or strategy deployment) is the result of integrating *kaizen* thinking into the annual planning process. The genesis of *hoshin kanri* (hereafter strategy deployment) is the blending of MBO, long-term thinking, and total quality management (TQM) practiced within leading Japanese companies in the

1960s. The stated aim of TQM was to build adaptive cultures. A typical Japanese TQM mission statement, in translation, reads: "Build organizational capability to change flexibly to the business environment through total focus on quality across all products, services, processes, and aspects of management by activating the potential of people."

The TQM approach lent scientific rigor to the MBO process. Total quality management was the overall umbrella under which customer-focused business excellence efforts were guided. Under this umbrella (Fig. 5.2) were total quality control (TQC) activities represented by quality control (QC) circles conducting team-based improvement activity, strategy deployment to drive the breakthrough improvements in the annual business plan across all departments, and what is simply called daily management, has been completely ignored in the West until very recently, when it was illuminated by Mann (2010) and others.

Adopted by Toyota, strategy deployment both gave structure to and became infused with *kaizen* values. At its simplest, strategy deployment

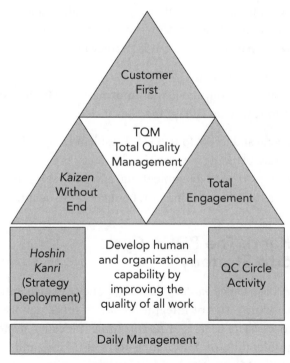

Figure 5.2 *Hoshin kanri* within TQM.

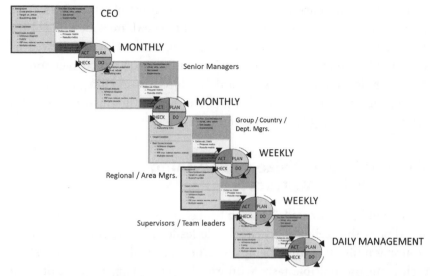

CEO

MONTHLY

Senior Managers

MONTHLY

Group / Country /
Dept. Mgrs.

WEEKLY

Regional / Area Mgrs.

WEEKLY

Supervisors / Team leaders

DAILY MANAGEMENT

Figure 5.3 *Hoshin kanri* alignment, consensus building, and follow-up.

should be considered "long-cycle PDCA," wherein breakthrough improvement targets are achieved through detailed annual planning, execution, review, and learning. The process and form of strategy deployment, although flexible and not an exercise in filling out a particular style of forms, is as important as the content. The process emerges from deeply held *kaizen* beliefs such as aligning the organization with long-term purpose, customer focus, fact-based understanding of one's processes, developing consensus, connecting activities at each level to the overall goal, and making plans tangible and possible to act on with urgency (Fig. 5.3).

Strategy deployment begins by identifying a small number (typically three to five) breakthrough objectives for the year. This act of limiting the high-level objectives from many to a few by itself can be a humbling exercise in honesty and understanding of the organization's true capability to implement its plans. Once the organization has defined what needs to get done, strategy deployment also helps to shape culture by defining *how* things need to get done. The first is to ensure coordination and connection across the vertical silos and functions of the organization so that everyone is working on the shared goals together. The "catch-ball" process of dialogue between leaders and subordinates at all levels converts the high-level goals into agreed-on and reasonable action plans, with an emphasis on understanding the current situation through data and observed facts. Once the plan is put into action, early and frequent reviews are done using visual management tools that act as early-warning signs. This is followed by structured problem-solving steps taken with urgency to correct course. Periodic reflection of progress toward the objectives and integration of the lessons learned into new standards drive the continuous-improvement activity of the strategy-deployment process itself.

The Vital Few: *Kaizen* Everything or Just What Matters?

If, as Franklin Covey said, "Human beings are wired to do only one thing at a time with excellence," then why do leadership teams believe that by grouping multiple human beings together they will be able to somehow each focus on multiple tasks with excellence? Perhaps because of what Kahneman (2011) has called "loss aversion," the idea that the fear of missing an opportunity is greater than the anticipation of gaining one.

The hardest thing about making executable strategic plans is deciding what we will not add to our list to do this year. The breakthrough objectives must be large enough to involve everyone, and when they are kept to a few, this allows the leadership to spend enough quality time in the check and act steps of the PDCA cycle. When there are tens of high-level objectives, key performance indicators (KPIs), or initiatives, simply reporting and reviewing the status take over as the main activity, with no time for detailed root cause analysis of missed targets and fact-based discussions of countermeasures. Leadership must bring not only direction but also focus to strategy. When there are many existing initiatives, sometimes it can be more productive to begin with a blank-slate vision exercise to identify the breakthrough objectives. These should be based on the deeply important long-term purpose of the organization. Humble reflection and a focus on understanding our customers, as well as understanding our processes and how well they are serving us, will provide focus.

Much like Moltke's expedients, frequently it is best to test the plan against reality—internal and external customers. Jeff Kaas, second-generation president of custom furniture and aerospace upholstery manufacturer Kaas Tailored, shares how *kaizen* has changed his approach to annual business planning:

Our planning method used to be to try to figure out what would happen in three years with the economy, government policies, [and] technologies and mold the organization accordingly. As in, if *X* happens, then we should grow by *Y*, and so forth. In hindsight, all of the planning used to be in batch. And if we had targets based on bad assumptions, we didn't change our plans midyear even if that didn't make any sense, because we didn't want to feel that we had failed. We have a statement of purpose that is updated once per year; it is just tweaked. We have a few long-term goals that everyone in the company is working on, a maximum of three at any one time. At the moment we are working on:

1. Cutting *muri* by 50 percent
2. Tripling the value of our routines
3. True pull

As a leader, I believe that causing or allowing burden and stress on the people is a moral issue. I want everyone in the organization working on cutting out half of the *muri*—overburden on people and processes—by the end of the year. I know [that] if we do these things, the business results will follow. Tripling the value of our routines means taking it two levels higher in terms of the discipline and ease with which we perform the basics—the daily, weekly, monthly, quarterly routines—things that need to get done. True pull has to start with leadership, with all processes having clear signals, as in "When you see this, act." One of the biggest changes has been that now our business plans are pulling ideas from the shop floor. Our goals are not top-down push, but pull. Three of nine of the initiatives under my top three annual objectives came from the shop floor. They understand their processes best and can tell the leaders what projects will most easily meet our goals. As a result, our *hoshin* boards are becoming real—red means red, *kaizen* actions are on the mark, these systems designed to draw action are doing just that. We are weeding, planting, and harvesting—focus of leadership is on pulling the weeds, allowing the harvesting to occur. We do not need a batch planning process. Now it's having faith that the business will grow if we pull the weeds; the more we pull weeds, the more room there is to grow.

Although we may say the customer is always right, we cannot stop there. We must ask one more "Why?" to find higher purpose and meaning. *Kaizen* seeks greater and greater good, which requires more than growth through customer satisfaction. This requires extending the value stream beyond the enterprise, or end-to-end supply chain, further upstream. This must include the natural cycle of raw material renewal and environmental stewardship. Indeed, it must extend all the way through to the customer experience, beyond the commercial transaction and into the impact that the proliferation of the product or service has on society across future generations. *Kaizen* simply limited to making better products faster and cheaper is uninteresting. It is not inspiring or motivating. It can be, in fact, stifling. A *kaizen* culture demands that we respect and elevate human potential to the highest possible level, and we find our higher selves when contemplating ever higher and more meaningful goals.

Catch Ball: Organizational Alignment the *Kaizen* Way

The example from Kaas Tailored of targets being set not only top-down but also bottom-up based on alignment and understanding of a few, large overriding objectives shows us a unique approach to catch ball. Catch ball is the two-way communication processes of strategy deployment to take high-level objectives and turn them into concrete actions based on the experience and wisdom of the people who actually do the work. In large, complex organizations, the process of going across as well as up and down to develop practical plans can take some time. However, if we believe in the *kaizen* belief of respect for people, understanding our processes and customers thoroughly and building consensus on the plan in order to act with certainty and urgency, we will find time for catch ball. It will save much time spent in correction later. Here are some key points to remember during the catch-ball process:

▲ Go to see the real situation to understand what can be achieved compared with what is being asked for. In some areas, what the people can deliver may be less than what the leader is asking for, whereas in others the people can deliver more. In both cases, it is safe to assume that the leader does not understand the process as well as the person closest to it.

Table 5.1 Linking *Kaizen* to Strategy at Each Level

Linking	Strategy	*Kaizen* Events	Daily
Scope	Business-level, enterprise value stream	Value stream, process level	Process, individual work area
Review Cycle	Annual, quarterly	Monthly, weekly	Daily, hourly
Teams	Management	Project, cross-functional	Natural work teams
Impact	Large, strategic, slow	Large, focused, fast	Smaller, focused, fast
Sustainability	Requires commitment to PDCA and constancy of purpose	Requires integration within daily and support *kaizen* structures	Requires team structure with team leaders as coaches

▲ Adapt the plan and adjust timelines, resources, and the balance of targets based on the overall situation and individual local realities, but do not compromise on the overall target.

▲ Use the three cycles of *kaizen* together to create additional capacity for innovation and strategic project activity. There is a "strategic gearing" effect (Table 5.1) when support *kaizen* (strategy), project *kaizen* (events), and daily *kaizen* are all done together, with greater impact than the sum of its parts.

Becoming quicker and more accurate with the catch-ball process must be one of the process metrics for learning about and getting better at strategy deployment, a topic of reflection at the annual review and planning cycle.

Alignment Between Process and Results at Lockheed Martin

As noted several times above, increasingly in the United States and around the world, federal departments, states, counties, and municipalities are recognizing that adopting operational excellence is a must to continue providing services in the face of limited tax revenues. This has been in no small part thanks to the influence of private industry and the aerospace and defense companies in particular with their direct exposure to the U.S.

Department of Defense. In the private sector, the emergence of *kaizen* culture has been more visible and rapid, especially among the largest organizations, which have been pursuing this path for decades. Lockheed Martin is an American aerospace and defense giant known for delivery of high-performance, high-quality products while lowering costs and delivering innovation. Senior Vice President Michael Joyce shares insight into the importance of alignment between *kaizen* activities on the front lines and the needs of the overall business:

> Every employee at Lockheed Martin knows that you've got to do your job, and you've also got to improve how your job gets done. The "how" behind this is something we call Structured Improvement Activity, or SIA, which is based roughly 80 percent on lean and 20 percent on Six Sigma. You can say "we need to do an SIA on this" anywhere in Lockheed Martin, and people know exactly what that means. Then, we've got the structure and tools to go do it. It's essentially a *kaizen* event.
>
> We've found this to be a very powerful system for removing waste and creating world-class lines. However, for a program like this to sustain itself over long periods of time, it has to work in cohesion with the way the overall business is being managed. The main key to that is making sure that you deliver outcomes. That's what the corporation measures do, and that's what we're paid to do.
>
> What *kaizen* pushes you to do is control the process, or the input, in order to get predictable output. That connection, however, is not easily appreciated by many financial managers. They tend to look at quarterly and monthly figures and come up with simple explanations for why things occurred. It's a lot harder for them to accept the idea of doing the heavy lifting to actually understand the root cause. So these are the kinds of tensions we have to deal with.
>
> I learned back in the '90s from the Toyota people that managing by quarterly, monthly, or even weekly numbers is not enough—you have to get down to a plan that measures expectations in hours. This means [that] if you're not on plan at 8:00, you can be problem solving at 8:01, as opposed to waiting

for the monthly review and then trying to figure out why you missed your plan. At that point, your problem solving is awful because it's not based in fact anymore—it's based in myth.

People working on the process side can lose sight of outcomes as well, and we sometimes have to remind them, "Are you getting predictable outcomes, yes or no? Or is *kaizen* your hobby?" Outcomes are the grade of the kind of *kaizen* culture you really have.

So you have to have alignment between your finances and your processes, and that's not a simple alignment to achieve. People say in three to five years you can do it in a lean transformation. That's wishful thinking. My belief is that it takes careers to make this happen, and even then, you're always going to have tensions between input-based and output-based decision-making.

Joyce alerts us to a long-term challenge that must be recognized and addressed as a strategic priority: the alignment between financial results and the processes employed to achieve them. By the nature of how financial results are measured, reported up through an organization, and presented to the shareholders within large and especially publicly traded organizations, there is ample opportunity for misalignment and disconnection between front-line actions and bottom-line results. There is indeed no easy short-term answer to this, although lean accounting models can bridge the gap to a certain extent.

Teaching *kaizen* practitioners who are process-oriented to see the outcomes of their actions, and teaching finance people who are results-oriented to look and and understand the day-to-day inputs are important first steps in finding common ground. Turning this potential area of misalignment and misunderstanding into a positive tension that stimulates discussion and problem solving is a key responsibility of the enlightened leader in shaping a *kaizen* culture.

From Operational *Kaizen* to Strategic *Kaizen*

For many organizations, the first obvious *kaizen* projects will be of the "correction" type, where processes are redesigned to solve problems in safety, productivity, quality, sales, inventory management, on-time delivery, or cost. During the daily practice of *kaizen*, people's eyes will increasingly take note of smaller problems or even potential problems and risks that are not yet today causing trouble, shifting the focus of activity toward prevention. Strategic *kaizen* activities beyond the six-month horizon or depending on the cycle of new-product introduction, new-customer acquisition, or new-process development should focus on anticipating and planning for unmet or unspoken customer needs.

Most of the *kaizen* work we hear about involves the practical deployment of existing strategies in the *gemba*—the actual workplace. *Kaizen*, however, also can be a powerful enabler for the formulation of the strategies themselves. Here Bruno Fabiano, director of the Kaizen Institute in Italy, explains how he helps businesses to create growth strategies using *kaizen* principles and methods:

> The *gemba* for senior managers encompasses all aspects of their company—the team members, the competitive positioning of products, the various customer segments, and the supply chain. Management teams also represent diverse competency areas, such as manufacturing, finance, HR, and sales. Leaders within excellent organizations are able to face the challenge to apply *kaizen* to every type of *gemba*.
>
> We begin the growth strategy creation process by helping the management team get a clear picture of everything that is happening in their *gemba*. To do this, we use two visualization tools from a methodology called Blue Ocean Strategy. One is the Business Model Canvas (Osterwalder and Pigneur 2010). This helps to make visual the company situation according to the nine attributes of:
>
> 1. Customer segments
> 2. Value propositions
> 3. Channels
> 4. Customer relationships
> 5. Key activities
> 6. Key resources
> 7. Key partners
> 8. Cost structure
> 9. Revenue streams

The second is the Value Curve or Value Profile, which creates a graphical representation of the positioning of the company relative to competitors according to factors such as price, customer service, and specific product attributes.

These visual tools are used to create maps first of a company's current state, and then, building on that, of the desired future state. The latter could include improvements such as expansion of product offerings, acquisition of new technology, alignment of costs according to competitive pressures, or the addition of new sales channels. The gaps between the two are used to create action plans that become the focus for *kaizen* improvement (Fig. 5.4).

Although these visualization tools are commonly used by traditional management consultants, our approach is very different in that we don't base our work on industry benchmarks, purported best practices, or other outside information. Instead, we use the *kaizen* principle of *going to the gemba* as our primary information source.

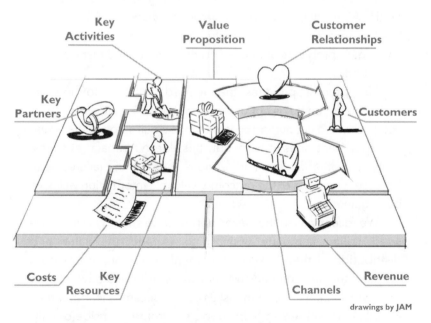

drawings by JAM

Figure 5.4 Business model canvas.
(Creative Commons.)

There are two aspects to this. First, we emphasize the wisdom of the team. We begin by telling them that "Nobody understands your business environment better than you." Secondly, when the team doesn't have all the data that is needed for a key decision, we give them the tools to extract that information from their *gemba*. This is often done with controlled experiments that test promising ideas in a low-cost, low-risk manner, in the same way that we use *kaizens* on the shop floor to conduct experiments testing new production processes. We do the same for business ideas and business plans.

Kaizen represents a paradigm shift for managers, and our first step is to change the way they think about their jobs. Even in many lean companies, managers are used to being given a strategy and spending their time figuring out how to deploy it. Here, their job is to work with the team to continuously improve the strategy. The metaphor is if you want to see, learn how to act. This is the way of thinking that we ask them to put into practice

Our process for designing and improving action plans is based on the A3 model, which was developed by Toyota. A3 is a step-by-step process based on the PDCA cycle where teams investigate problems or needs and then create and test plans for resolution.

The process provides the needed structure for low-risk experimentation but is not overly complex, which allows teams to act quickly. We encourage clients to take small steps so that when an experiment doesn't work out, this doesn't impact revenues, costs, and other factors. Furthermore, there are no failures here, because regardless of the outcome, valuable information will have been gained.

We recently used the *kaizen* approach to strategy with a mid-sized Italian company wishing to expand the sales and distribution of their bakery equipment. The mapping process revealed that the company was not well-positioned beyond their domestic markets. Through discussion, it became clear that they would need partners to help support customers outside of Italy, partly because of the need of their customers to have 24/7 support.

The A3 process was used to test different relationships with potential partners and find the most advantageous arrangement. The result was a successful expansion of sales channels, which allowed the company to increase sales volume at a time when their market was actually declining.

A similar approach was used with a mid-sized furniture provider seeking to improve their sales in the United States. They had previously shipped only full containers from Italy to customers who had large enough orders. This cut out many potential customers.

There were several options available to provide a distribution model suitable for smaller orders. For example, they could rent a warehouse, buy a warehouse, or contract out the warehousing of their goods. What they needed was data to make a proper decision.

To gather these data, they conducted an experiment. They implemented a six-month trial arrangement with a dealer. Under the agreement, the dealer provided the required warehouse space and was given a special discount, which allowed him to serve customers with smaller orders within a defined territory. This gave the company a practical way to test the viability of the smaller-order market with solid evidence from their own *gemba*.

Of course, this kind of experimenting is outside of the comfort zone for many managers who are accustomed to using market data and other conventional sources. The main value of using the *kaizen* approach to develop and implement growth strategies is that we help these managers break through this barrier and learn to trust their data and trust the wisdom of their team members. The key is to take small steps that prove what is possible and act as a leverage point for changing their views.

Once these barriers are overcome, it is surprising how quickly teams can put these ideas into practice. In the two cases above, managers were calling dealers and signing agreements within a matter of hours. Once they start to move, there's nothing that can stop them.

This is a powerful illustration that there is not any particular tool, consulting framework, or business-school paradigm that results in a successful growth strategy. Rather, it is the culture of an organization that is entrepreneurial and able to take intelligent risks. In the preceding examples, the risks were small, easy to test, and showed success or failure quickly, allowing learning and adaptation. The *kaizen* approach as strategy in practice enables the management team to use the scientific method while engaging their people to work toward a common purpose. The impact on the organizational culture is to build trust, and the financial impact is positive both on the top and bottom lines.

No Plan Goes According to Plan

There is a powerful saying that is part of many Eastern philosophical traditions: "Plans don't always work out the way you imagined, but they always work out the way you implemented." This could not be truer of business plans and business strategies. The value of applying *kaizen* values and methods to the process of strategy development and execution is that it reinforces the habitual and frequent turning of the PDCA cycle at all levels in the organization. This frequent check catches small deviations from plan and exposes inaccurate assumptions that we made when setting our annual plans and, most important, the performance-eroding behaviors that arise from deeply embedded beliefs within our culture, such as:

- ▲ Silent disagreement
- ▲ Agreeing to high goals knowing that they will be adjusted downward later
- ▲ Agreeing to high goals knowing they will be lowered later
- ▲ Not challenging the leader's plans to avoid damaging one's career
- ▲ "Making the numbers" by adjusting outputs, knowing that the bosses will not check

Author, speaker, and Professor Bob Emiliani of the Center for Lean Business Management is an outspoken critic of continuous-improvement efforts that are misapplied. His "Dooming cycle" parodies the Deming cycle (a.k.a. the PDCA cycle) but strike at a deep truth about how many of us manage (Fig. 5.5). Organizations that do not establish systems to make problems visible at the earliest stage, invest in turning every person into a

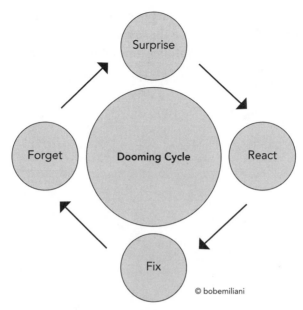

Figure 5.5 The Dooming cycle.
(© Bob Emiliani. Used with permission.)

problem solver, insist on taking root-cause countermeasures, and learning from both success and failure are doomed to the vicious cycle of surprise, react, fix, and forget only to be surprised again, reacting anew. This behavior of coping with problems as they arise rather than planning ahead to prevent them is called "firefighting."

Following the *kaizen* process develops learning organizations that do not forget the lessons of past problem solving. Such organizations are rarely surprised by the same problems recurring, and are adaptive and able to respond when surprised by new challenges. The learning organization not only recovers from challenges but adds the latest lesson to their organizational memory bank. Embedding these habits into the organization requires strong leadership to follow the PDCA cycle for all strategy, business planning, innovation projects, and problem solving. When top management insists on this *kaizen* process from the strategic level through to the front lines, this results in learning at all levels and allows escape from the dreaded Dooming cycle. The building of habits, the catching of problems at their earliest and smallest, the building of practical problem solving skills, the safeguarding of standards that enable consistent performance, and the giving of live feedback on where the organization is going off course on strategy—these are all in the too-often-unexplored domain of daily *kaizen*.

CHAPTER 6

Daily *Kaizen*

> The secret of getting ahead is getting started. The secret of getting started is breaking your complex overwhelming tasks into small manageable tasks, and then starting on the first one.
>
> —MARK TWAIN

It has taken almost 30 years of staring at the obvious, but increasingly people are realizing that the key to sustaining the benefits of implementing any management system or system of business excellence is the day-to-day, hour-by-hour management practices that leaders make a priority. On the surface, the content of *kaizen*-style daily management appears boring, even oppressive, to professionals and leaders who value the freedom and autonomy that their position gives them. Daily management involves going out of the nice office to the noisy front lines to check deviations from standards, observe internal and external customer-supplier interactions, and ensure that corrective actions are put in place and standards are updated. These are the many small, urgent things that are essential to sustaining high performance. The traditional management approach, characterized by management by objectives (MBO), command and control, batch and queue, and leave your brains at the door, will struggle to see the sense of paying so much attention to front-line people and processes. An entirely new way of thinking is required, one that can only be learned by doing.

We call it *daily kaizen* because although the management practices and skills are exactly those involved in other cycles of *kaizen*, be it of the project type or strategic *kaizens*, the focus is on maintaining and improving performance within each day. On a daily basis, *kaizen* builds skill and confidence and delivers business results by encouraging people and teams to take repeated small steps forward. These steps are so small that it is nearly impossible to fail in a big way, and with supervisors and managers nearby as coaches and supporters, small failures turn into big learning. The combination of many small successes and perhaps a few big ones, plus continuous learning, builds confidence in and commitment to the new system. This type of on-the-job training and coaching requires different skills from the front-line leader, who frequently is simply the most senior person, with little formal management training. Therefore, daily *kaizen* requires that we identify and develop formal and informal leaders.

In daily *kaizen*, the benefits of *kaizen* activity are seen not as either financial results or people development but inherently both. The daily cycle of improvement reinforces the learning and development through practice of problem solving while adding small financial benefits that add up to large benefits over time. *Kaizen* projects and breakthrough events deliver larger results more quickly. Daily *kaizen* enables people to learn how to work within the newly designed process day to day while adapting to challenges that come from situations that the process was not designed to handle, thereby learning problem solving skills. The natural teams on the front lines become the safeguard against backsliding from *kaizen* events. At the support *kaizen* level, leadership steers and guides the prioritization of breakthrough events to achieve business goals while learning through follow-up and daily *kaizen* how these step improvements can succeed in practice. Leaders, in turn, learn how to shape the organization, developing new skills and making new connections between people in order to make high performance sustainable. Before we concern ourselves with sustaining the gains, we need improvements, and before we have improvements to sustain, we must have standards.

Maintenance of Standards

In his groundbreaking follow-up book, *Gemba Kaizen*, Imai (2012) observed that successful day-to-day management could be reduced down to one concise precept: Maintain and improve standards. Perhaps as a result of the fact that readers were drawn to the allure of *kaizen* and less or not at all to the more mundane activity of *maintenance*, the most important half of sustaining the gains of *kaizen* has been largely neglected until recently. Let's take the analogy of owning a high-performance personal vehicle. Let's assume that such a vehicle is more fun to drive, accelerate, map the journey, steer the course, tune up, or even refuel often than an average vehicle. But most of us who are not trained mechanics would not choose to maintain an expensive high-performance vehicle ourselves and certainly not to tune it up on a daily basis. Without maintenance, before long there would be no vehicle to drive. At best, we would spend our time and resources on neither maintenance nor improvement but rather on repair. Because we are busy in repair mode, we have little time for upkeep and no time for improvement. In the modern organization, we are all passengers in the same vehicle, and we all must pay attention to maintenance of the vehicle and improvement of the journey. Daily *kaizen* turns this activity of maintenance and improvement into a virtuous cycle that develops people and processes while delivering results (Fig. 6.1).

There are three critical components to the daily-management precept of maintain and improve standards—standards, maintenance, and *kaizen*. As we showed in Chapter 2, the emergence of a strong foundation of standards within the best companies in Japan (which is not to say all companies) after World War II was directly related to the quality improvement efforts initiated by the U.S. occupation forces and the American quality experts they brought in to help rebuild Japanese industry. The teaching of statistical control methods and the scientific approach to

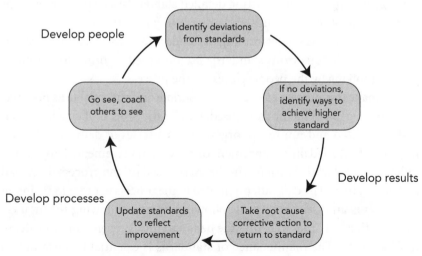

Figure 6.1 Virtuous cycle of daily management.

improvement embodied in the plan-do-check-act (PDCA) cycle underlie the existence and acceptance of a generally more robust level of standards. However, standards are not simply things such as product specifications, technical settings, or the amount of time needed to perform a task. More broadly, management must understand standards to mean the normal and correct way things are done, rules, and even performance targets toward which to strive. Imai described these key features of standards:

1. Represent the best, easiest, and safest way to do a job.
2. Offer the best way to preserve know-how and expertise.
3. Provide a way to measure performance.
4. Show the relationship between cause and effect.
5. Provide a basis for both maintenance and improvement.
6. Provide objectives, and indicate training goals.
7. Provide a basis for training.
8. Create a basis for audit or diagnosis.
9. Provide a means for preventing recurrence of errors and for minimizing variability.

The maintenance of standards, the first of these core daily management activities, is not as mechanical and labor-intensive as the analogy of the high-performance vehicle used earlier might make it sound. Within a *kaizen* culture, the expectation of management is not that a significant part of each day is spent updating or rewriting detailed standards. The primary actions of management to maintain standards are to go see the process in action, check that the process is performing to standard, and use Socratic questioning when deviations from standards are observed in order to stimulate problem solving by the people closest to the process.

To make this process of checking as simple and sustainable as possible, visual management methods are used. Finally, the process of checking itself is standardized and improved, organized as a series of daily, weekly, and monthly checks—daily management or routine management. This is often called *standard work for leaders* in the West based on the process described by Mann (2010). The daily attention paid to the process in order to maintain standards is an essential part of following up and following through on *kaizen* activity in order to realize the benefits (this is covered in more detail in Chapter 8). The maintenance of standards is essential to sustaining a *kaizen* culture, and this requires management to create an environment that is safe, stable, and free from excess burden and to connect people in natural work groups or teams.

How Much Time Should a Leader Spend on the *Gemba*?

The best answer to the question, "How much time should a leader spend on the *gemba*?" is to reflect on the question, "How much of your business do you want to be blind to?" The *gemba* is where the people, information, processes, and customers meet in ways that have the potential to create value or generate non-value-added activity. It is the source of truth, and like truth, if grasped should set us free. The goal is not to spend hours standing on the front lines looking at processes and trying to find waste, except as an educational or eye-opening exercise or as part of a periodic mental recalibration. The role of leaders is to make the direction and targets clear and visible so that ideas for achieving those targets can come from the team.

This is illustrated in the examples from Medtronic in Chapters 3 and 7. Although it is possible to give various guidelines such as, "For supervisors, 80 percent of the time on the front lines; for general managers, 50 percent," and so forth, this is a meaningless exercise until the processes are understood, the teams are defined, the team leaders are trained, visual standards are put in place, minimal daily routines and norms are established, and at least one concrete way to engage in *kaizen* is put in place. Only then will the front-line people and processes create a true "pull" for support and leaders will know when, where, and how much they are needed. There is no harm in selecting an area as a model area and applying a formula for time allocation toward daily routines, as found in Mann (2010), but this should be done as one pilot project with the primary aim of learning and adapting to one's unique environment. For the senior leader, it is recommended that he or she spend as much time as possible on the *gemba* until the purpose of this behavior has been understood by all subordinates, and they have taken concrete action to make reality visible on a daily and hourly basis.

Every Day Improvement

We have said that the true ideal of *kaizen* is realized when everyone is engaged in making bad things good and good things better, everywhere, every day. Eiji Toyoda, former chairman of Toyota, said, "Our workers provide 1.5 million suggestions per year, and 95 percent of them are put into practical use." Most of these ideas are small, and the stated purpose of everyday improvement at Toyota can be seen on the signs around and within the company's facilities: "Good thinking, good products." The workers at Toyota implement one or more improvement ideas per person per month. As worldwide employment at Toyota exceeded 250,000 people, the total number of improvements implemented also grew. Furthermore, each year there are literally millions of conversations taking place to develop small observations into full-blown and implemented *kaizen* ideas.

Kaizen is not inherently a daily process for humans. Both thinking and making improvements require energy. Making small improvements every day goes against our nature, which is to conserve energy, stock up on food and other vital supplies, avoid exposure and risks, and protect our own status and security. The "smart ones" and survivors scheme, bide their time, and make big improvements. Everyone wants to get rich quick, to look for the "big score." Slow and steady may win the race, but we grow anxious as we feel the pace of daily life speeding up.

There are deeply wired behaviors that prevent us from looking objectively at ourselves, analyzing root causes of our behaviors, and taking many small steps to improve, all the while scientifically validating or invalidating the results. Instead, we jump to solutions, justify our positions emotionally, and move on to the next and bigger problem when the current one is contained to a reasonable level or at least swept under the proverbial rug. In order for everyday *kaizen* to become a reality, leaders must have a sense of urgency about it. Some may say that a quota such as "one *kaizen* idea per person per month" is oppressive, but it is in fact no different from any other requirement of employment. We must make our way to our place of work at least one time per day that we are employed. (Perhaps for many of us this is the harder action than finding one simple *kaizen* idea.) Avoiding the minimum requirement for improvement activity is an arbitrary abandonment of management responsibility. It shows that we have not yet embraced the *kaizen* beliefs of alignment with a long-term purpose to succeed by developing people and serving customers, having a sense of urgency and respect for people, and taking time to understand our processes together.

Organizations that fail to insist on deliberate practice of daily *kaizen* by everyone put themselves at risk of backsliding. They will experience the gradual erosion of any benefits achieved through both daily improvements and massive improvement projects or strategic investments. From the point of view of benefits, daily, bottom-up *kaizen* activity:

▲ Promotes creative problem solving
▲ Shows respect for people
▲ Increases employee engagement
▲ Speeds up the detection of problems
▲ Reduces costs as a result of the preceding

Despite these benefits, one of the main reasons that daily *kaizen* is interrupted or stopped is that from a financial standpoint, it can be hard to justify allowing individuals to spend 10, 20, or 30 minutes several times each week engaged in problem solving activity. Although immediate attention to costs is necessary, this must be balanced against the larger goal of aligning and engaging everyone in the urgent task of improving every day. We have learned through conversations with management during over 100 study missions to Japanese companies that when *kaizen* suggestions are part of daily management systems, they yield savings of 300,000 Japanese yen per person per year, on average, approximately $250 per month. Aggregated and averaged across the organization, suddenly these ideas do not seem so small.

Growing People Every Day at FastCap

FastCap is a product-development company based in Bellingham, Washington. Established in 1997 by Paul Akers, a veteran woodworker and inventor, FastCap engages its entire workforce of 56 people in the maintenance and improvement of standards every day. There are several notable features to the *kaizen* culture at FastCap. The company culture would not be what it is without the unique passion, urgency, and energy of Paul Akers as its leader. As an inventor, Akers is never satisfied, always finding something in the world that does not work and inventing a solution. This makes Akers a demanding leader, a tireless teacher, and an inexhaustible source of ideas. When Akers discovered *kaizen*, he found a way to help everyone in his organization see the processes in his company in the same way that he saw the words *always something to improve*. In order to make it as simple as possible for his people to understand and immediately take action, Akers demanded that everyone make a two-second improvement every day. He wanted to engage every person's mind at FastCap in finding something that could be improved, no matter how small. Akers even wrote a 100-page book about this approach entitled, *2 Second Lean: How to Grow People and Build a Fun Lean Culture*, to share the story. Akers commitment to developing his people into problem solvers who are aligned with his vision and who understand the process takes the form of the daily morning meeting. Every day for 30 minutes to an hour, all 56 employees of FastCap stand in a circle in the middle of the facility (Fig. 6.2). They review the mistakes that were made on the previous day, the reason why, and what can be done to prevent similar mistakes in the future. They engage in peer-to-peer teaching, rotating a different employee as the teacher each day. They watch a video to learn about history and discuss its lessons. They start their day as one team.

This time costs the company $2,000 per day, according to Akers. A simplistic cost calculation would show that this daily "nonproductive time" across 220 working days equals nearly a half million dollars per year. How can a small company possibly afford to do this? Akers asks, "How can you possibly afford not to?" FastCap ships a high volume of orders, and we have no doubt that the two seconds of savings per person per day easily add up to far more than the cost of the morning meeting. Although *kaizen* is an improvement to the standard and thus a self-funding activity, this is not the right way to calculate the cost and benefit of daily *kaizen*. There are many

Figure 6.2 The daily morning meeting at FastCap.

indirect benefits of the *kaizen* culture that lower costs and accelerate growth at FastCap but are much harder to calculate, such as:

▲ A positive, fun, engaged workforce in a family atmosphere
▲ Prevention of errors, avoidance of repeated errors, a quality mind-set
▲ Ease of gaining top talent by becoming an employer of choice in the region
▲ Development of future leaders with deep knowledge of the company
▲ Increasing international exposure for FastCap and its products through excellence in *kaizen*

No doubt many readers will remain in shock and dismiss out of hand the possibility of having the entire production team, much less the entire company, including its president, meet every day for 30 to 60 minutes for learning and improvement. Yet how many hours each day do people in your organization spend in meetings in which zero learning and zero improvement happen? How far away from the actual workplace do these meetings take place? How fresh or accurate is the information being discussed at these meetings? How many times was the information transferred from point of occurrence to the reporting point at the meeting? What causes such meetings to be necessary?

If we view the daily *kaizen* activity at FastCap as a countermeasure to certain management challenges and reflect on the situation within our own organizations, we can find the answer. In the broadest sense, we can view the daily morning meeting at FastCap as the ultimate act of maintenance of standards. Each morning the entire workforce confirms its commitment to the company, to each other through the history lessons and peer-to-peer teaching, to the customers through quality reviews, and to their many shared cultural values. The cost of trying to manage and repair broken processes that would result from stopping the daily maintenance would be far greater than taking one hour every day to align and energize the team.

Team Development Program at Sonae Retail

Sometimes the spark that begins a *kaizen* transformation comes from unexpected places. This was the case with Sonae, a European retail chain employing more than 25,000 people with revenues of more than 3 billion euros. In 2006, a government mandate for all companies to provide a minimum of 35 hours of vocational training per year to employees took effect as part of a European Union strategy to increase competitiveness. Sonae had to comply, but it needed to do so in a way that met a key requirement—not to interfere with activities on the retail floor or disturb supermarket customers in any way. The productivity-improvement training had to blend seamlessly into the daily work. The approach the company designed with the help of the Kaizen Institute was called the "Team Development Program" (Imai 2012).

Launched with strong leadership support from the CEO and COO, the program began with management training to build a strong guiding coalition that could communicate what *kaizen* meant for Sonae. The first step was to define the natural teams and then to engage the managers, supervisors, and employees in these areas. To make the activities sustainable, the targeted areas for daily *kaizen* activity were kept highly visual and practical, including:

▲ Training Within Industry (TWI) instruction to help workers integrate *kaizen* into their daily routine
▲ Standardized job instructions that enabled people to work without checking with management constantly
▲ Improvements to make self-maintenance of equipment possible

This was a bottom-up effort, with training delivered one *kaizen* theme at a time for two to four hours per month. The main objectives were to create standards and make problem solving visual.

Based on the training, included in the daily routines of the natural teams were activities such as

▲ A five-minute shift-start meeting to check team performance metrics, record and review problems, celebrate success stories, and propose improvements
▲ On-the-spot problem solving using a visual contain, control, and correct (3C) approach
▲ Upkeep of safety standards, 5S workplace organization, and housekeeping

Although daily *kaizen* was a bottom-up activity at Sonae, the Team Development Program itself was a support *kaizen* activity led by management. The program was organized as part of an annual training plan, with

Figure 6.3 Daily *kaizen* in Sonae retail stores.

agendas, training manuals, promotion plans, and audit plans to check effectiveness and improvement. A small number of internal trainers led the training while informal and formal opinion leaders were nurtured to create behavior tipping points and reinforce the daily *kaizen* routines (Fig. 6.3).

Through the Team Development Program, daily *kaizen* activity was cascaded to more than 20,000 employees over a period of five years and continues today as an organic part of the company management system. Although the initial aim was productivity improvement, the daily *kaizen* activity changed the culture in noticeable ways. With the insights that "The best way to learn is to teach" and "If the student hasn't learned, the teacher hasn't taught," the human capital-development approach was energized like never before. Given the history of success and growth over 25 years, the culture at Sonae was characterized by the belief that the current way was the right way. Typical words of resistance to change were, "We have always worked this way" and "I have no time for this type of thing." However, as time went by and people saw the results, even the most vocal critics turned into the most enthusiastic supporters.

Nowadays Sonae is developing the IOW (Improving Our Work) program which integrates the efficiency and continuous improvement methodologies in all company areas. It also promotes the growth and expansion of the continuous improvement culture.

What's So Hard About Simple *Kaizen*?

Getting the process right to make small, simple improvements looks very simple but is actually very challenging. Steve Jobs (*BusinessWeek* 1998) said, "That's been one of my mantras—focus and simplicity. Simple can be harder than complex: You have to work hard to get your thinking clean to make it simple. But it's worth it in the end because once you get there, you can move mountains." As an example, the short-cycle *kaizen* activity known as the creative idea suggestion system has been in place since 1951, when it was adopted from the model at the Ford Motor Company. However, as we explained in Chapter 2, the daily *kaizen* activity at Toyota was by no means an immediate success. It took the leadership of Shoichiro Toyoda to organize and direct the daily *kaizen* activity (Wakamatsu 2007). After he became the head of Toyota's suggestion system committee in 1967, he reviewed the award and recognition system, established factory-based committees to promote *kaizen* suggestions, and worked to increase the quantity of ideas generated by employees. During his tenure as head of the committee, the number of *kaizen* suggestions increased by 600 percent. Further adjustments were made to the program over the years to improve both the quality and quantity of *kaizen* ideas.

The suggestion system is one of the standards that requires constant maintenance, upkeep, follow-up, and promotion to sustain. Designed properly, it is more than self-funding from the savings generated by the *kaizen* ideas. The value of safety and quality problems prevented, improved engagement and morale, and increased levels of problem solving and presentation skills is difficult to quantify financially, but such activities are even more significant. The guiding principles for the design of a system for simple daily *kaizen* include the following:

▲ Integrate improvement into everyday work.
▲ Focus scope of ideas on small, local changes that the natural team can make.

Figure 6.4 *Kaizen* wall of fame.

▲ Give direction to improvement ideas by addressing gaps in performance targets.
▲ Make improving one's work part of everyone's job.
▲ Make helping people improve their work part of a leader's job.
▲ Review ideas informally, verbally, and on the spot, and document after implementing.
▲ Recognize, reward, motivate, and promote individuals and teams (Fig. 6.4).

There are some assumptions underlying these design principles. As we will see in Chapter 9, for a positive transformation of any type to succeed, certain organizational readiness actions must be completed. For example, the scope of the *kaizen* suggestions is kept small enough to decide through discussion with the team member and team leader, ideally arriving at actions they can implement or at least try without seeking higher approval or budget. This presupposes that there is a team and a team leader with enough knowledge of the work and capability to evaluate improvement suggestions, as well as the time, authority, and support to make the changes that are sensible. For small, informal organizations, this is not a large hurdle; it is just a decision. For larger, more complex organizations, this requires various reviews and realignments of policies and responsibilities.

Putting in place a *kaizen* suggestion is in itself a *kaizen* project. To those who would say, "It's too much work! Let's just do large projects and *kaizen* events," we would ask, "What percentage of your people do you want to exclude from the maintenance and improvement of standards?" Total engagement requires daily *kaizen*—tangible daily targets to strive for, things people can change, and experiments to engage their minds. Most of the brains in the majority of organizations reside not in managers, engineers, or professional staff but rather in front-line workers and supervisors. It follows that most of the improvement ideas will come from nonmanagers, nonengineers, and nonprofessionals. Furthermore, the closeness of front-line workers to customers and the process means that they often have information that is better and fresher than anything that managers at a distance from the *gemba* could hope to access on an hourly basis. However, paradigm-shifting improvements and innovative ideas often come not from natural teams that work closely together each day but from bringing cross-functional teams together for focused sessions under the guidance of a skilled facilitator. Given time and focus to gain a deeper understanding of the process and the requirements of customers, given challenging targets to stretch people's thinking and comfort level, and given the authority to make physical or system designs that literally change the scenery, *kaizen* project teams achieve breakthrough results. Equally important in creating an an adaptive, human-centered and high performance culture, such *kaizen* teams change mind-sets, grow emotionally, and build belief. We will study how this is done in Chapter 7.

CHAPTER 7

The Emotional Lives of *Kaizen* Teams

A man without a smiling face must not open a shop.

—CHINESE PROVERB

The ability to persuade people to change is a critical skill of any leader. The leader who leads based on authority alone can only go so far. Such a leader can only be right so many times, and when he or she is wrong, often a culture of fear prevents people from facing the leader's error and correcting course quickly. Authority-based leadership, while appropriate in some situations, is fragile. The approach that Toyota has taken to leading product development is instructive in terms of the value of leadership by persuasion rather than authority. Every Toyota vehicle has what is known as a *shusa*, or "chief engineer," responsible for the concept through launch and market entry. This person has responsibility but not authority.

How does this work in practice? Can an engineer with no authority influence the decisions and direction of development of new products, new technologies, and new methods of manufacturing and distribution? The answer lies within certain deeply anchored values at Toyota of making the progress of often complex research and development processes visible using the *obeya* "open-room" method, the value placed in learning from successes and failures via practice of the plan-do-check-act (PDCA) cycle, and respect for individuals. The last is put into practice by the art and skill of persuasion that the chief engineer must possess. Persuasion is much more than merely having all the facts, knowledge, ideas, and logical arguments to make a case. The leader who is able to get people to follow is persuasive through a combination of these things in addition to character, credibility, and understanding of human psychology and the role of emotions in learning and decision-making.

Ethos, Pathos, and Logos

Greek philosopher Aristotle wrote more than two millennia ago that the secret of being a persuasive speaker lies in the effective combination of ethos, pathos, and logos. *Ethos* is an attempt to persuade by establishing the character, credibility, or authority of the speaker. *Pathos* is an appeal to the emotions of the audience, to persuade through an emotional connection. *Logos* is persuasion logic. Together these three are called the *persuasive appeals,* and they apply equally to the process of implementing rapid change as they do to persuasive speaking. All three appeals are an important consideration in shaping a *kaizen* culture, but within organizations, the most neglected and of greatest importance is the appeal to pathos.

Although not every *kaizen* leader may possess the right balance of credibility, ability to connect emotionally, and ability to persuade their colleagues logically to go along with the changes proposed, it is crucial that he or she possess at least two of the three. A credible leader who is able to connect emotionally will still be able to persuade the team to try the changes. By following the *kaizen* process together, they will be able to arrive scientifically at the most logical proposal. The *kaizen* leader who lacks ethos or credibility will still be able to succeed when he or she is able to make emotional and logical connections. This is why even very young people can be successful as *kaizen* consultants if they possess these two persuasive appeals, whereas very senior consultants who possess ethos and logos but not pathos can struggle when lack of human empathy trumps authority and logic.

The emotional connection to the process of change, the leaders and team members going through the change, and how individual integrate this into their experience and worldview are very important for adult learners. Unlike children, who can learn new things quickly, adults require more time to rewire their brain, reset habits, and change patterns of behavior. Studies have shown that this is a painful process, both physically and mentally. It does become easier with practice and under guidance from an emotionally intelligent *kaizen* leader.

In his early days as a consultant, one of the authors, Jon Miller, made the "rookie mistake" of relying on logos more than pathos. As a consultant in his twenties, he made up for what he lacked in credibility that comes from years of experience with a thorough knowledge of *kaizen* principles gained from working closely with Japanese masters. During a project to redesign the assembly and packing line of a small cabinet manufacturer, Morgan, the plant manager, had doubts about removing the conveyor and shortening the line. These doubts persisted even after review of the results of the time study, line balancing, proposed combination of steps and removal of non-value-added motions, and the resulting productivity calculations. The author's words were literally, "How can it not work? It's science." From the expression on the plant manager's face, the words might as well have been, "Are you stupid?" Partially reassured that the experiment was reversible, the new design was a great success, and it was the first of many *kaizen* projects for this plant manager and coworkers.

Why was there reluctance about making what was on paper a simple and easily reversible change to the assembly and packing line? No doubt part of the hesitation was due to the inadequate level of ethos or credibility established by the consultant at this point because it was his first project at this company. The logos, or logic, was airtight, but the pathos, or emotional persuasiveness, was lacking. One out of three of the persuasive appeals was at least one too few.

The role of pathos in *kaizen* is even more important than its role in the persuasive speaking that Aristotle wrote about. In *kaizen*, we are not debating or asking people to change their beliefs or positions; we are asking to change physical or transactional processes that have immediate and real-life consequences to the organization. The benefit to the company from a successful change may be great in terms of higher quality and lower costs, but the benefit to the individual middle manager may be smaller, perhaps recognition in the form of a year-end performance bonus or a future promotion. On the other hand, the risk for the company is a failed experiment that can be reversed with no permanent damage. For the middle manager, on the other hand, the failure could result in a variety of personal setbacks, both real and imagined. Persuading potential blockers of improvements requires attention to the emotional connection, often the fear of negative consequences more than the desire for gain.

Kahneman (2011) and others have shown that people tend to strongly prefer avoiding losses to acquiring gains. This is called *loss aversion* and explains why there is resistance to change from middle managers. Studies suggest that losses have twice the psychological impact when making decisions than gains. When shaping a *kaizen* culture, it is necessary to make explicit that blame is not assigned for making mistakes but that learning from mistakes is expected and failures are tolerated as a natural consequence of experimentation and improvement. Even when everything is done right, we cannot avoid the fact that going through major changes of the types experienced by organizations embracing *kaizen* in the pursuit of service excellence can require a literal rewriting of the neurons in our brains, a change of habits and routines, and a reassessment of how we think about ourselves in relation to our work and our colleagues. When the changes presented pose a threat to one's authority, power, or position, the emotional reactions can be strong.

Helping People Scale the Change Curve

The five stages of grief introduced by Elisabeth Kübler-Ross are a good model for understanding the reaction people have when faced with changes that involve actual or perceived loss (Fig. 7.1). Psychologist, author, and grief counselor, Dr. Kübler-Ross (2005) found that people go through these stages in different ways at different speeds. At first, there is typically denial that the change is happening or that it will affect oneself, then anger as the inevitability dawns on them, then bargaining in an effort to find a way to

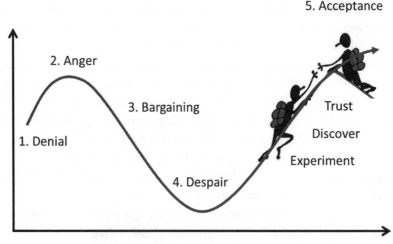

Figure 7.1 Scaling the change curve.

reduce the loss, and then depression as a feeling of powerlessness sets in. The key for leaders guiding change is to help people scale the change curve from depression to acceptance as quickly and smoothly as possible. Fortunately, unlike the types of losses Kübler-Ross wrote about, often the expected losses as a result of a *kaizen* transformation are not as bad and not permanent. In addition, *kaizen* itself empowers people going through the change to have a hand in it. Therefore, the time spent in stage 4 (depression) can be minimized by supporting and coaching people toward stage 5 (acceptance) through hands-on learning and experimentation within the new way of working.

Leaders must keep in mind the *kaizen* core beliefs of humility, security, alignment with long-term purpose, and respect for people when introducing a major change effort to the organization. This means having open and honest two-way communication about what will change, how people think the change will affect them and how it will in reality, what is known and what is not known, and how people can get involved in moving from stage 4 to stage 5. The emotions must be engaged at the same time as the intellect.

In many organizations with a masculine culture, the very first step may need to be simply making it safe to talk about one's emotional states. Emotions are things that we all have, we all show, and we should all do our best to master and improve. As the transformation effort progresses, people will need to let go of existing habits and create new habits and routines. Much of this change occurs on an emotional level and takes time, as with training a muscle. Whenever possible, this is best done in ways that are reinforced by the day-to-day responsibilities that are accomplished within natural work teams. This does not mean that we need to make *kaizen* projects into emotional events to discuss our emotions. In some cases, the *kaizen* event process is useful in bringing out old resentments in a constructive way. For example, frequently people and processes that are disconnected grow to silently blame the other but lack the avenue to have an open, customer-focused problem solving discussion. Physical and functional walls do not help in this regard, but in a cross-functional team environment, where titles are left at the door and everyone works toward the same goal, people quickly find that they can move from emotion to objective discussion.

Many animals, including humans, have natural visual controls to signal emotions. People can rarely hide how they feel, whether through body language, facial expressions, or tone of voice. A common facilitation tool is

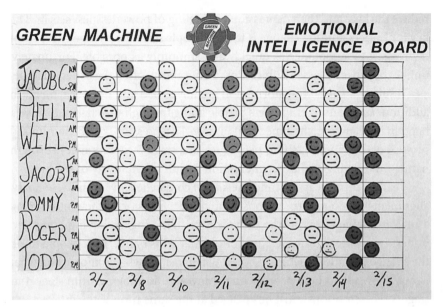

Figure 7.2 The emotion meter.

the emotion meter (Fig. 7.2), which can be posted, encouraging people to use a picture or drawing to show whether they are having a neutral, bad, or good day. This does not excuse bad behavior, but it allows us to respect each other, to recognize that we all have good and bad days, and to support each other during the *kaizen* event.

As difficult as change can be, managing change via a *kaizen* approach can have great personal benefits as well. Maurer (2004) found that it was possible for people to make dramatic personal change by starting very small and building bit by bit, in effect, removing excuses of "no time, no energy" and so forth and recognizing the power that people do have to change something, even if it is small. Maurer's research showed that the smallest acts of caring, the smallest efforts to exercise, or meditate, and small inspirations led to better marriages, better health, and greater productivity and innovation. We remove fear and build habits by experimenting and discovering that we can, and by doing so we incrementally build the trust and confidence that we can persist where in the past we gave up. As adults at work, we can't always intervene in the lives of other adults to help them overcome personal problems. This may be the function of the extended family and friends, the community, and religious groups, but it is not a primary function of the workplace. However, it is not only compassionate but a rational business decision to take an interest in people and their lives and to do what we can to equip them to improve their lives. These same problem solving and communication skills will make them better employees and better future leaders at work. Even people who are perfectly content have dreams and ambitions, and *kaizen* skills can be applied to target setting as well as problem solving, to helping people plan and achieve their goals. There is no downside in this regard.

Given the rapid and focused nature of the process of facing problems, understanding the process elements that cause them, testing out changes, and learning as a team, the *kaizen* event is a unique opportunity to help people to quickly scale the change curve. Indeed, any type of intensive workshop or boot camp that is professionally designed and facilitated with this aim can guide people through the change curve safely. Managed poorly, this can create resistance to change or even a strong dislike of *kaizen* or change of any kind. Managed well, it can be a source of commitment, energy, and belief in where the organization is going.

The Triple Purpose of a *Kaizen* Event

Although we often speak of *kaizen* events serving a dual purpose—to deliver business results through concrete change and to develop the problem-solving skills and other management skills of people—there is a third factor. When *kaizen* events are conducted with the 10 core *kaizen* beliefs in mind and the company's explicitly stated principles are kept in front of people, this has the effect of anchoring them in belief through experience. Examples of such experiences include

▲ Going in person to see the real situation in order to *understand the processes.*

▲ Exposing problems within a *secure, blame-free environment.*

▲ *Connecting people and processes* within the company and understanding what it means to work together as a team.

▲ Seeing many ideas being put into place with urgency, within hours or days.

▲ Witnessing management commitment to the *consensus* plan through follow-up and support to see the *kaizen* ideas put into action.

▲ Seeing that leadership is investing time and money during the *kaizen* event in the *long-term purpose to succeed through developing people.*

▲ Experiencing a sense of *humility* and *respect for people* on hearing normally quiet or unappreciated people speaking up with great ideas and insights into the process.

Although there is a well-tested format for preparing for, facilitating, and following up after *kaizen* events, these actions should not become formulaic or mechanical. The appropriate balance of results, intellectual learning, and emotional growth will not be the same every time. *Kaizen* is human-centered scientific problem solving that uses focus, team effort, and speed to help people scale the change curve and integrate their emotional responses to the process of change.

Emmanuel Dujarric, senior director of Medtronic surgical technologies at Medtronic and leader of the factory in Jacksonville, Florida, has a clear understanding of the importance of helping people to see the meaning of *kaizen* activities and how they align with long-term purpose. In the main meeting room, once a week, leaders get together to review how effective *kaizen* is supporting Medtronic's mission and goals by using magnets cards on the wall. The Medtronic mission is broken down into categories that are linked to key indicators for each of the value streams in the plant. Then *kaizen* projects or events are agreed on to achieve the goals, and their status is reviewed weekly (Fig. 7.3).

When people can link their *kaizen* efforts to the very reason why they work at Medtronic, not only will the company avoid wasting its limited

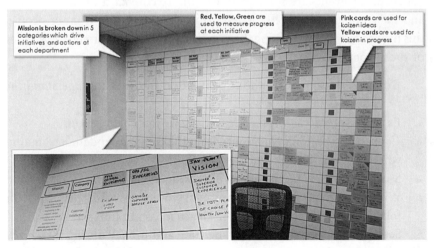

Figure 7.3 Linking *kaizen* activities to purpose at Medtronic.

available resources, but people also will be motivated to complete and collaborate to achieve company goals. This builds emotional as well as intellectual engagement.

Creating Emotional Engagement in Kaizen at Wiremold

Perhaps no other experience stirs up such a variety of strong emotions among the people in an organization as a company merger or acquisition. Changes that result can include company name and identity, leadership team, employment policies, and incentive programs, even employment itself. On the one hand, people feel anxious and unsure about the future, whereas on the other hand, people feel hope and excitement at the opportunities that may open up under new management. A *kaizen* event is a small change relative to a company undergoing a merger or acquisition, yet we find that *kaizen* can enable and make this process more effective when leaders grasp the emotional landscape of change.

Few people, if any, have overseen the introduction of *kaizen* into more companies than former Wiremold CEO Arthur Byrne, now applying his experience with *kaizen* to the venture-capital and private-equity fields. He recently authored *The Lean Turnaround* (McGraw-Hill, 2012), in which he recounted his experiences and shared keys to success. After transforming Wiremold through *kaizen*, raising productivity by more than 150 percent, gross profit by more than 13 percent, and operating income more than 13-fold, Byrne led similar transformations across more than 30 companies in 14 countries. Here he explains his process for leading a company through that critical phase when facing great change:

> When I was CEO of Wiremold, we acquired a number of companies and introduced *kaizen* to all of them. The approach we always took was to first get everybody in a room. After introducing myself and Wiremold, I would tell all the employees where we were going, why we were doing it, and the kinds of results we expected.
>
> I also outlined the kinds of changes employees were going to see, and why this would be good for them. I said right off the bat that we were not going to lay anybody off as a result of this *kaizen* work, because the initial response of most employees when you're

introducing something like this is to be scared stiff that this is just a way to cut jobs or some other kind of thing.

The other key was to make people understand that they were going to have a big input into the how. This wasn't something that off-site consultants were going to do to them. The employees themselves would be the ones doing this through a series of *kaizens*, over a long period of time.

Next, I personally conducted a three-hour overview of lean and *kaizen*. The main idea here was to get everybody on the same page, and to show that we were serious about doing this. Having the CEO do the presentation is, I think, critically important to get that particular message across.

We followed up that introduction with the first *kaizen* on the very same day. We'd tell our new associates, "Let's start with the doing. We will explain it to you up front. But then, this afternoon, we are going to start doing *kaizen*. And we are going to attack a couple of areas, and we are going to make progress this week. And it is going to look very different in these two areas than it did on Monday when we showed up."

By the afternoon, we were moving equipment around and people were saying "Who are these guys?" But, on the other hand, there was never any question or doubt whether we were serious, or what we wanted to do. By the time they went home on the first day, they knew where we were going.

So people really didn't have a chance to catch their breath. But, on the other hand, the results that we were demonstrating after a day or two were already spreading around the place, and people were saying, "Wow, this is unbelievable what they're doing. It's so much better."

Arthur Byrne sets a high bar for CEOs in personally leading a company through *kaizen* during a critical phase. Each of us must decide what level of success we are committed to achieving and how closely we can follow Byrne's example. We can extrapolate these general principles of *kaizen* leadership from his experience:

1. Personally communicate what will change and why.
2. Create psychological security by committing to no layoffs due to *kaizen*.

3. Give people control over their destiny—the *how*—of the long-term change via *kaizen*.
4. Deliver three hours of education (by the senior leader).
5. Get to work that same day and deliver "Wow" results.

The first thing a leader will need in attempting this approach is a lot of help. Few CEOs understand what is required to kick off a successful *kaizen* event on the first day, much less stay through the entire three- to five-day workshop. Moreover, for all *kaizen* events, but especially those at critical phases such as in the preceding example, little is left to chance, and much preparation goes into making sure that the team has everything they need to make "Wow" changes from day one. CEOs have the power, resources, and skill to get this done, but this requires will.

Changing the Scenery Within the Week

The difference with daily *kaizen* or management-level strategic *kaizen* activity is that *kaizen* events involve cross-functional teams, have projects with plan-do-check-act (PDCA) cycles compressed into five days rather than across weeks, and have a bias toward rapidly redesigning and implementing processes. Significant visible change happens within the week. There is a Japanese expression often used in the context of *kaizen*: "To change people's mind-sets, first change the scenery" or, perhaps simply, "Seeing is believing." We can talk all we want, but there is nothing like learning by doing, experimentation with new ideas, and moving aside furniture or equipment that has been there for years. These actions enable people who are at the bottom of the change curve and feeling helpless to gain a sense of empowerment. These are not symbolic actions or merely exercises in rearranging the furniture. The changes follow the principles of creating flow, leveling and balancing workload, signaling work from the customer pull, and building in quality in each step. When such practices are already in place, the opportunity for such "change the scenery" *kaizen* events becomes much less common, as has been the situation at Toyota for many years.

Common ways to change the scenery include visiting a world-class benchmark company to see how others have made a positive impact with *kaizen*; doing a small demonstration activity such as 5S to clean up and reorganize a work area to improve safety, quality, and productivity; or completely redesigning the workflow or work method in an area within a

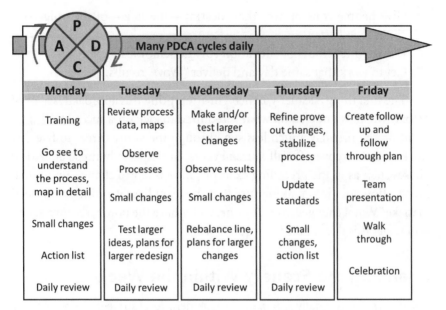

Monday	Tuesday	Wednesday	Thursday	Friday
Training	Review process data, maps	Make and/or test larger changes	Refine prove out changes, stabilize process	Create follow up and follow through plan
Go see to understand the process, map in detail	Observe Processes	Observe results		Team presentation
	Small changes	Small changes	Update standards	
Small changes	Test larger ideas, plans for larger redesign	Rebalance line, plans for larger changes	Small changes, action list	Walk through
Action list				
				Celebration
Daily review	Daily review	Daily review	Daily review	

Figure 7.4 Typical *kaizen* event day by day.

week. Seeing is better than hearing, and doing is better than seeing. The *kaizen* event both changes the surroundings of team members and puts them in an environment that is completely different and can achieve this change in mind-set. The changes often happen literally overnight; physical or system changes proposed on Tuesday may be completed by Wednesday morning (Fig. 7.4). Often there are support resources on standby to respond with urgency to help the team realize its ideas. For people who are not on the *kaizen* team but work in the area, being shown a redesigned area can bring about an emotional reaction that can be excitement, enthusiasm, surprise, shock, embarrassment, or even denial. It is important to involve and communicate with these people as much as possible so that they are moving along the change curve at a similar pace as the *kaizen* team responsible for redesigning the process during the week.

The *Kaizen* Team Presentation Day

In the wrong hands, the *kaizen* team presentation on the last day of a *kaizen* event can appear to be a brief show of results to justify the next consulting engagement, continuation of the budget for such activities, or even "ticking the box" to fulfill a corporate mandate for number of *kaizen* events held per year. This final stop on the emotional journey of a *kaizen* event team is much more than just a show and tell. It is customary for the cross-functional *kaizen* team to include few, if any, managers and certainly not as a majority of the team, except in cases where the theme of the *kaizen* project requires it or where it is organized as a learning event for a group of managers. *Kaizen* team members are expected to be dedicated to the team for the duration of the event, and this can be difficult for managers in organizations less advanced in shaping the *kaizen* culture, where firefighting, meetings, and surprises are daily events. In addition, managers have duties that require travel to customers, suppliers, or other divisions of the company, making it difficult for them to stay with a team for four to five consecutive days. For these reasons, it may be the first time that some members of the management team see the team in action. It is important that managers know what to look for and how to behave.

First, it is necessary to recognize that in many cases the *kaizen* event team members are not practiced presenters. Many people have a dread of public speaking, and here, after a long week of hard work, each team member is asked to stand in front of management to speak for a minute or two about his or her experience or some element of the *kaizen* event week. Emotional states among the team members that must be respected include

▲ Fear of public speaking
▲ Pride at accomplishment
▲ Disappointment at not being able to finish all the actions on their list
▲ Hope for the future
▲ Concern that there will not be support to implement or sustain
▲ … and many others

The leaders listening in the audience must pay attention to more than just the words but also to the body language, tone of voice, and evidence of the emotional lives of the *kaizen* event team members during this presentation. Depending on the occasion, it may be appropriate to encourage team members to verbalize their concerns, ask how they will communicate their positive feelings to infect others with the enthusiasm for *kaizen*, and make an emotional connection by committing forcefully to seeing their efforts through to completion. Such acts of emotional intelligence are not always the competencies that have made leaders successful within an organization, but they are the skills necessary to shape a *kaizen* culture, and the *kaizen* presentation is a safe venue in which to practice and hone these skills.

The role of the managers or leaders who are listening to the *kaizen* team presentation includes

▲ Recognizing the efforts and contributions of the team
▲ Recognizing members individually
▲ Asking questions about the process and what people learned
▲ Asking questions about the results and where they see future opportunities
▲ Reinforcing any messages needed to remove fear and build engagement

The *kaizen* event presentation is also an opportunity to reinforce behaviors and practices that support the shaping of the organization's culture. A partial list of evidence to look for and offer recognition for includes

▲ The team's ability to organize the problem solving story logically
▲ The ability of team members to present their ideas clearly
▲ The use of simple visuals
▲ Correct use of problem solving tools and understanding of the scientific method
▲ Teamwork between the members of the *kaizen* event during the presentation itself
▲ Time management—the use of a time keeper—giving each presenter a set amount of time
▲ Appropriate humor, camaraderie, and team identity—are they having fun?

The manager who is capable of coaching team members in these areas is encouraged to be present during the *kaizen* team rehearsal, typically an hour or two before the actual presentation, in order to encourage team members and give them pointers. As time and resources allow, *kaizen* team leaders and even team members should be given practical training in these skills. The behaviors in evidence during a well-executed *kaizen* team presentation are all characteristics desired within an organization's culture. The *kaizen* event is a controlled environment within which to practice and improve these skills, but it should not be the only environment. Supervisors and area managers who lead natural teams in daily *kaizen* also should be trained in these skills so that they become behavioral norms, part of the culture. It should be noted that if the routine management meeting were run with the structure, professionalism, and preparation of the *kaizen* event presentation, the productivity of the management staff would soar to unprecedented levels, freeing these knowledge workers to pursue innovation, new markets, and talent development.

Every *kaizen* event that ends well increases the chances of the next *kaizen* event succeeding. Building the desire among people to participate in *kaizen* events is an important factor. This requires making sure that it was a satisfying experience, both professionally and emotionally. A team presentation that is well attended by management, where team members tell their story and together the organization commits to seeing the team's effort through, is key to a *kaizen* event ending well and *kaizen* itself continuing.

The Facilitator's Role in Leading People Through Kaizen at ThedaCare

As director of education at ThedaCare Center for Healthcare Value, Marta Karlov has participated in hundreds of personal learning journeys through *kaizen*. Here Marta discusses the role of facilitators, coaches, and mentors in navigating the deeply personal inner process of change:

> I was trained as an industrial engineer, so when I first started leading *kaizens*, the process part was ingrained in my DNA, and I just assumed this was easy for everybody. When people told me I was taking the steps too quickly for them, this helped me become much more aware of all the different ways that people perceive their world.
>
> We use a compressed Plan Do Check Act cycle in our *kaizen* events, and over the four days, things can get pretty emotional. At the beginning, there's a lot of brainstorming, and people saying "This is good; we can do this," but once people start to see the current state, and how fraught it is with waste and chaos, feathers start to get ruffled. It's hard not to take it personally when the team decides that something you were responsible for is not delivering the results.
>
> Being a good facilitator requires a certain level of self-confidence and ability to lead a team. When I coached facilitators,

I did a lot of observing followed by real-time feedback. Sometimes I would show them first, and then help them practice, with lots of positive reinforcement.

The trick here at least as far as I'm concerned is to meet people where they are. Really understand who they are, how they learn, what drives them. This really goes back to what I learned at the beginning—that everybody learns differently, and everybody sees the world differently. Part of this is that nobody has all the answers, and people have to be allowed to fail.

The facilitator's job is to teach people the PDCA cycle so they can test and learn from it, not to get in front of them and say, "Hey, that's not going to work." If people fail, that's okay, and they might surprise you.

I think that before change can take place, trust has to be built—trust that the leader will follow through, that the followers are important, that their opinions are important, that team members have the leader's backing. Once you build that trust, it's amazing what change you can bring about.

The key lessons for facilitators to effectively guide teams through their emotional lives during a *kaizen* event are to:

1. Respect people as individuals with different learning styles and reactions to change
2. Create a safe environment for people to learn from failure
3. Build trust as a leader by listening and following through on people's ideas

The facilitator's role within a *kaizen* culture involves much more than guiding people through the change process; it includes helping people to grow and the organization to develop by anchoring the changes as positive emotional experiences.

A Kaizen *Event with Shigeo Shingo at Hill-Rom Industries*

In July 1985, Mike Wroblewski was still a junior industrial engineer in the Fabrication Department of Hill-Rom Industries. The company had invited

Shigeo Shingo, a consultant to Toyota and author of 10 books on industrial engineering, *kaizen*, and the Toyota Production System, to teach the single-minute exchange-of-dies (SMED) system. After Shingo met with the management team, Wroblewski, a setup operator and a tool-room technician, was selected to work with Shingo for one week to understand how the SMED system worked. Wroblewski did not know who Shingo was at the time. Shingo explained through an interpreter that we should be able to exchange our dies in 10 minutes or less. Working in the Fabrication Department, Wroblewski knew that the setups and die changes took between one and four hours on the presses that ranged from 75 to 400 tons. He could not conceive how this could be done in less than 10 minutes.

The *kaizen* team watched and videotaped a typical die change, which took an hour. Shingo explained the difference between internal time and external time, asking, "What things can we do while the press is still running the previous job to have things prepared ahead of time?" He wanted the team to move activities that were internal to the downtime to be done external to the downtime—while the machine was still running—in order to reduce the die change time. The team followed this guidance and changed its method, resulting in a die change that was cut to 30 minutes. "We thought that was great, the best ever," recalls Wroblewski.

But Shingo was not satisfied. "Please try again," he said, asking that the team look at standardizations and adjustments. The team noted that the die heights were all different, took measurements, and made the die heights and pass lines the same. As a result, the 30-minute changeover time was cut to 20 minutes. Team members were delighted. But still, no celebration. "Please try again," said Shingo, and this time he asked the team to look at bolts and clamping, which they were able to reduce from 18 different bolts to a few slotted attachments. As the team practiced the changeover, dies going in and out, team members improved little things each time across the week. The final result was that a changeover that took over an hour on Monday was performed in less than 5 minutes on Friday. The team had beaten the 10-minute mark.

"Shingo made us feel that we did it ourselves and he just pointed the way. It was the greatest feeling in the world to us to accomplish what was thought to be the impossible. My world of possibility and improvement became endless," said Wroblewski. Indeed, during the week, the team made the emotional journey from shock and denial ("It's not possible") to acceptance ("Maybe we can do it") to a feeling of pride at achievement. Thus it was a surprise at the end of the presentation of results when Shingo said, "Please try again." He gave further hints on how the team could still save 5 seconds here, 10 seconds there. The message sunk in: Continuous improvement truly is continuous, and there is always something more to improve.

"What we learned that day," says Wroblewski, "is that although all week we thought the 10-minute mark was the goal, it was only the direction."

Wroblewski and the team members experienced many emotions, no doubt positive and negative, during their *kaizen* event with Shingo. Also, it is clear that there were times when they were confused about what was possible, what they must do, and finally, what the long-term goal of the SMED exercise really was. This confusion was resolved by the very end, resulting in an "A-ha!" moment for them.

A study by D'Mello and Graesser (2012) suggests that negative emotions and "productive confusion" can help students to learn more effectively in complex learning situations. Participants in the study reported being in neutral emotional states about a quarter of the time while feeling surprise, delight, engagement, confusion, boredom, and frustration during three-quarters of the time. D'Mello and Graesser found that learners move back and forth between states of "engagement/flow" and "cognitive disequilibrium" as they are faced with obstacles or contradictions and they are able to resolve them through thought, reflection, and problem solving. The uneasy feeling of being mentally thrown off-balance motivates us to find the answer, and learning occurs.

This is significant for people going through *kaizen* transformations because they will be faced with many counterintuitive experiences while practicing *kaizen*. These ideas include the fact that the vast majority of the work we do does not add value, that less inventory is often better, that one-piece flow is more productive than batching, that checking at every step costs less than checking quality once at the end, that it is better to stop work when a problem is found than to keep working, that taking time each morning to meet as a team results in better performance than saving that time and using it to start work early, that the smallest changes can have big impacts, and so forth. The mindful *kaizen* facilitator can even use emotions such as confusion to the advantage of the learning process.

Why Toyota Doesn't Do *Kaizen* Events But You Should

The *kaizen* event is not practiced at Toyota in the popular one-week, cross-functional team format. The closest type of *kaizen* activity at Toyota is the *jishuken* workshop, but this is aimed primarily at developing managers, is shorter in duration, and does not typically involve major process redesigns. The big changes are done as technical *kaizen* projects, without a cross-functional team, involving line workers for a week. Some use this fact to argue that the *kaizen* event is not a "pure" approach, as though Toyota claimed to be the source of truth. Toyota arrived at its *kaizen* culture in a different way from anyone else over a period of decades. The closest thing the company had to an external *kaizen sensei* to teach them was Shigeo Shingo, who was an industrial engineer and consultant brought in to help establish the SMED system at Toyota. As we explained in Chapter 2, many of the raw building blocks of the Toyota Production System were learned from American teachers and shaped into the Toyota Production System through trial and error.

The *kaizen* event is certainly not the best or sole way to achieve culture change, but it is an essential accelerator in allowing people to build belief in the methods of *kaizen*, trust in the behavior changes of leadership, and importantly, believe in the opportunity to emotionally integrate large changes in the process, physical environment, and methods—what a *kaizen* team goes through day by day during a typical one-week *kaizen* event as team members repeatedly turn the PDCA cycle.

Many managers who have left Toyota to go to work at another company as the internal lean *sensei* first experience culture shock because the basic assumptions of management are so different. In fact, these *sensei* are at a disadvantage in many ways, like a fish out of water, until they are able to realize that they have left the pond. Many have never experienced a *kaizen* event or equivalent, having learned the Toyota Production System from great teachers but in a much more measured, much less emotionally taxing way. With an enlightened and committed leadership team willing to give the ex-Toyota lean *sensei* the time to build a *kaizen* culture gradually and level by level without cross-functional team breakthrough events, they may succeed. Too often, needing fast results or being the one seen as the mad prophet making crazy claims, both sides part ways in disappointment. Luckily, the *kaizen* event format is not difficult to learn, and many people have added it to their list of skills. The ability to engage individuals and teams with emotional intelligence is a much deeper skill. It is necessary not only during the process of executing a change, such as a *kaizen* event, but well afterwards, during the all-important follow-up and follow-through phases of change.

Follow-Up and Follow-Through

What thoughts come into mind when you hear the words *follow-up* and *follow-through*? Ironically, "lack of" tops the list from our informal survey. We hear about great analysts, great strategists, and great doers but rarely do we hear about great follow-uppers. Why is this?

Let's start with the emotions that these ideas conjure. For many of us, follow-up leaves us with a feeling of incompletion, a job yet to do, a chore, an interruption, or a distraction from the urgent or interesting matter at hand. Having to follow up or having someone follow up can leave us feeling hesitance, reluctance, or even dread. Follow-up happens because something is not done, something is not done because there is a problem, and if we follow up, this means that it could become our problem. Cultural assumptions about problems, responsibility, urgency, and teamwork all play into how people follow up. When we explain the changing roles of leaders within a *kaizen* culture, often the reaction is, "Following up on strategic plans, okay. That's monthly reviews. This is part of management." When it comes to following up on *kaizen* events, the reaction is, "I can handle this. It's not like we have a *kaizen* event every week." Following up on daily maintenance and improvement activities on the front line, though, the response is, "Now that's going to take some effort!" In a *kaizen* culture, follow-up is seen not as a nuisance or a drag but something to get excited about, something to look forward to, part of the plan from the beginning. The follow-up plan can be simple and part of the standard routine that is developed in the course of linking daily *kaizen* with the management activity.

During a *kaizen* event, it is possible to accomplish much, but there will nearly always be additional improvement ideas and action items that must be completed in the following days and weeks. These items are listed on a *kaizen* newspaper (Fig. 7.5) that is then posted in a high-visibility location near the area where the *kaizen* event took place. The name *kaizen newspaper*

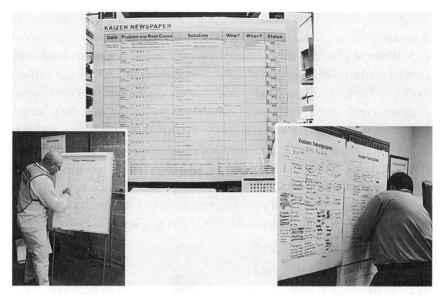

Figure 7.5 *Kaizen* newspaper.

comes from the fact that it is meant to be news, fresh information that is read every day. It is not an action list that is filed away and read only rarely but often a large piece of paper placed on a wall near the work area where the status of actions is highly visible.

Former Toyota Chairman Fujio Cho stated the *kaizen* spirit with regard to follow-up elegantly: "Go see, ask why, show respect." Go see also means to be seen following up. Although it is not "for show," it is important that people see and feel that follow-up is taking place. Following up and checking the *kaizen* newspaper after the front-line staff has left or before they arrive defeat the other two elements of Cho's purpose of following up. Showing respect means that checking is an act of caring about people and the common goals we are working toward, not an act of policing to catch people not completing tasks on the *kaizen* newspaper. The follow-up is also to see how the people are working within the new process, whether there are difficulties that need urgent attention, and by doing so building trust in the *kaizen* process.

Finally, to ask why means also showing humility, curiosity, and a willingness to help while at the same time checking the thinking process of the person being asked and providing coaching and development as necessary. Just as there is a check process built into the strategy deployment cycle, there is a built-in development purpose to the review and coaching activity within daily maintenance and improvement. Given the project nature of *kaizen* events, meaning that they are discontinuous events when compared with *hoshin*-type support *kaizen* or the routines of daily *kaizen*, *kaizen* events can become formulaic with the expectation that they are "over in five days." Leaders must be vigilant against this by following up in person until the new process developed during the *kaizen* event has been handed over successfully to the natural team for daily maintenance and improvement. Every project and every *kaizen* event must be followed by a reflection session in order to complete the full PDCA cycle, asking:

1. What was the target?
2. What was the actual result?
3. Was there a gap between target and actual (positive or negative)?
4. If so, why?
5. Did we follow our standard process?
6. If not, why not?
7. What countermeasures or updates do we need to make to our standard for *kaizen* events?

Follow-up is nothing short of an indication of character—of caring about the things we are responsible for. One of the best and most overlooked pieces of management advice can be found in Geissel (1971): "Unless someone like you cares a whole awful lot, nothing is going to get better. It's not."

From *Kaizen* Event Results to *Kaizen* Cultures

The fact that *kaizen* events bring quick results—often double-digit improvements, in safety, quality, time, and cost—has nothing to do with culture. *Kaizen* actions deliver results because poor processes are replaced with better ones, and good processes yield good results. It is thanks to scientific problem solving and the application of superior practices during a *kaizen* event that teams achieve breakthrough results. *Kaizen* is nothing if not practical. However, the sustainability of these results is all about culture, and this is often the major blind spot when attempting to apply the visible systems of lean manufacturing (a.k.a. the Toyota Production System) such as daily huddles, 5S, quality circles, or *kanban* systems. These are not systems in the sense that they operate when turned on, unmanned. These are human systems that rely on certain ways people are organized and connected to each other and behaviors and rules (often unwritten) of interaction. These behaviors must be underpinned by shared values and assumptions or people will drift apart from them.

CHAPTER 8

Sustaining a *Kaizen* Culture

Watch your thoughts, they become words; watch your words, they become actions; watch your actions, they become habits; watch your habits, they become your destiny.

—LAO TZU

Having a strong culture is not the same as having a *kaizen* culture. Often a visionary founder or dynamic CEO can set the tone for an organization, single-handedly shaping and sustaining the culture over years. Unless such an organization consciously decides to make the values explicit and link them with the long-term purpose of the business, there is a risk that a strong culture will not survive a leadership change. A *kaizen* culture not only must be strong, it also must be adaptive in order to survive a leadership transition.

By now the reader may realize that act of creating a *kaizen* culture cannot be separated from sustaining it. One goes with the other in a continuous cycle of reinforcement and adjustment. It is like walking. One cannot go far with only stretching one leg; at some point, we realize that we need the other leg before falling. However, many companies find the need to test how much they can postpone the sustaining part before falling to the "old ways" of doing business. It is not a coincidence that the number of organizations failing to create a lasting culture of improvement is probably less than 5 percent. By missing the sustaining part of creating a *kaizen* culture, not only is the effort invested so far lost, it also becomes harder to try again. People lose belief and motivation, and emotional resistance builds up to failing again. This behavior at the leadership level may be seen as a lack of commitment—"Here we go again"—with the demise of the program of the month. So how do we nurture a *kaizen* culture? Before we can answer this question, we need to understand and agree on the target condition and the problem that we have in front of us.

The Odds Against Survival

"It is not necessary to change. Survival is not mandatory." With these timeless words, W. Edwards Deming invited those wishing to survive and thrive to embrace change. Survival is the result of an organization's capability to adapt to change. Although high performance can be achieved through diverse means, the acid test comes with time. Sustainable high performance comes from nurturing a *kaizen* culture based on core beliefs and adaptive behaviors. Sustaining a *kaizen* culture is holding the line of defense to maintain the continuation of your success.

Foster and Kaplan (2001) shared two interesting findings about sustaining high performance. The first is about the extreme difficulty, or inability, of corporations to sustain high performance over time. These authors researched the 500 major U.S. companies within the Standard & Poor's (S&P) Index to learn about their performance over time. They identified the only 74 companies out of the 500 listed in 1957 that remained in the index by 1997. Fewer than 15 percent of the best and most successful U.S. companies were able to sustain their performance after 40 years. By 1998, only 12 companies were able to beat the S&P Index performance, fewer than 3 percent. The news gets even worse; the second finding shared by Foster and Kaplan is that not only is sustaining high performance difficult, it is also becoming harder than ever. They found that the average survival time of corporations in the S&P Index since its creation in 1920 has been reduced from 65 to 10 years by 1998. At this rate, Foster and Kaplan concluded that over the next quarter century, two-thirds of the major U.S. corporations will not be among the 500 major corporations. More than ever, a sustained *kaizen* culture with people-centered scientific problem solving is needed not only to achieve high performance but also to sustain it. To nurture an authentic, adaptive, and attractive culture increases the odds for survival and success by focusing on principles that drive both high performance and relentless innovation.

To Sustain Is to Change

Let's first agree on what it means to sustain in the context of *kaizen* culture. The term sustain has eight definitions in the online *Merriam-Webster's Dictionary*. The three first definitions are (1) "to give support or relief to," (2) "to supply with sustenance: nourish," and (3) "to keep up, prolong." Our meaning for *sustain* in the context of *kaizen* is closest to the second definition. To sustain a *kaizen* culture means to supply sustenance to develop people's thinking and behaviors with the objective to make continuous improvement a habit. To sustain a *kaizen* culture is therefore not about maintaining or supporting a status quo but to foster people to challenge it in a scientific and systematic way and to adapt it to current needs. The meaning of sustaining is a dynamic force, continuously changing to nurture *kaizen* culture creation. The meaning is not at all conserving, protecting, or unchanging, which is a common misunderstanding about sustaining a culture or system. A farmer does not simply plant seeds in the field, leave them to the forces of nature, and expect to harvest a crop in the future but rather constantly monitors the crops during their growth, provides fertilizer and water, and removes weeds and pests that threaten it. A farmer maintains and improves his or her fields. A farmer would take some seeds from the best crops and use them to plant again next year, thereby improving year on year.

The key action to sustain the benefits and processes of *kaizen* culture is daily practice. We keep up the ability to change and adapt by building up habits and routines. These routines, in turn, reinforce *kaizen* beliefs. This is accomplished by the continuous repetition of both the maintenance activity and the improvement activity of various plan-do-check-adjust (PDCA) cycles in combination. Whereas the maintenance cycle reinforces the behavior, the PDCA cycle improves it to a new higher level (Fig. 8.1). The continuous or unending repetition of both maintenance and PDCA is the

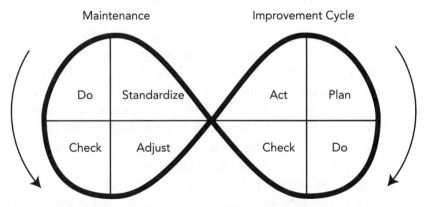

Figure 8.1 Maintenance and improvement cycles to sustain.

pursuit of perfection using a systematic scientific approach. In fact, Taiichi Ohno stressed that the PDCA improvement cycle cannot be achieved without the maintenance cycle (Ohno 2012).

The Toyota Production System uses the maintenance and improvement cycles very intensively. To the untrained eye, this may appear as unnecessary or even overly constraining, but routine develops the right habits through repetition. Take the example of an operator performing an inspection by taking a marker and lining around the edge of a part before he or she moves the part to the next operation. While marking the part is not necessary to perform the act of inspecting, what the operator is doing is making the inspection routine visible so he or she always starts at the same point, direction, and sequence.

According to the Centers for Disease Control and Prevention (CDC), about 50 percent of the 45 million Americans who smoke try to quit every year and fail. We all have experienced that same frustration of trying and failing to change even small habits. So how should we make *kaizen* principles stick in every team member's mind as the new way to do things?

Sustaining Through Habits

There is truth in the saying, "It is easier to *act* your way into a new way of thinking than to *think* your way into a new way of acting." If we could change how we act just by thinking, we would all think away bad habits and think up better habits. Yet changing bad habits is difficult. Making *kaizen* core beliefs and behaviors an intrinsic part of how we think and act means building the habit of continuous improvement. This is no different from any other habit, requiring us to apply the principles of habit formation. This calls for more than just reading any book about *kaizen*; it requires a focused and deliberate effort to create a new habit. We don't create good driving habits just by reading about driving techniques, lecturing people, or even memorizing a car owner's manuals. To stop smoking, understanding the bad effects of smoking helps, but usually is not enough for most of the 50 percent of Americans who try to quit every year, as we learned earlier. Therefore, we need to understand how habits are formed first.

Duhigg (2012) studied and wrote about the power of habits, explaining that neuroscientists have identified that habits are formed in a part of our

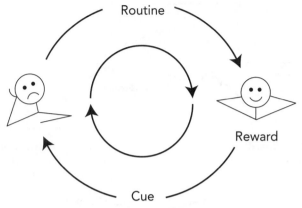

Figure 8.2 The habit loop.

brain linked to the development of emotions, memories, and pattern recognition, away from the prefrontal cortex where decisions are made. This specialization of our brain makes it possible to repeat learned routines without thinking about them; we are just acting in automatic mode. Duhigg indicates that to create any habit, three elements are required to create the "habit loop" (Fig. 8.2).

As an example, when a piece of work does not meet a quality standard, it is placed in a red bin (the cue); automatically, the supervisor takes the part out of the red bin and gathers the team to apply practical problem solving (routine) to identify the root causes. Here is where a *kaizen* practice such as root-cause analysis via the "five-why questioning" is applied. Satisfaction (reward) occurs when the root cause creating the out-of-standard part is quickly addressed, and the team meets its target. What we see in this example is that the three elements of habit formation as part of a system: Cue (red bin), routine (practical problem solving), and reward (achieving targets) are linked as integral parts of a system that reinforces *kaizen* thinking and doing (Fig. 8.3).

Rother (2010) calls these routines observed within Toyota *kata*, from the Japanese word for the form or way of doing something. Rother makes an important distinction between principles and *kata*. Whereas principles "help us to make a choice, a decision," *kata* "tell us how to do something, how to proceed, what steps to take." Sustaining a *kaizen* culture involves not only incorporating routines but also adding the right cue and reward to

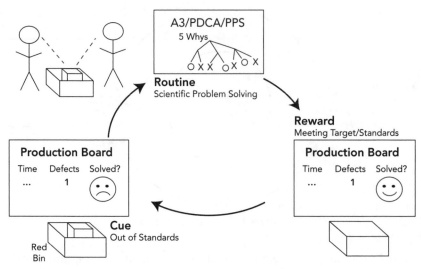

Figure 8.3 Building *kaizen* habits.

Table 8.1 *Kaizen* Habit Checklist

Cue	Routine	Reward
1. What is the cue?	4. What is the routine?	8. What is the reward?
2. Is it visually evident?	5. Is it clearly defined?	9. How effective is the reward to reinforce the routine?
3. Is it easily linked to the routine?	6. What *kaizen* principles are followed?	10. How is it linked to the cue?
	7. Is the routine effective?	

reinforce and turn the routines into habits, and this requires experimentation (Table 8.1).

Positive- and Negative-Reinforcement Loops

Ironically, the behavior of highly undesirable and contagious diseases such as bird flu holds the key for sustainability of *kaizen* culture. A particular flu strain can start in a remote village and in few days reach a metropolis with millions infected. Wouldn't we like to have a similar "*kaizen* culture infection

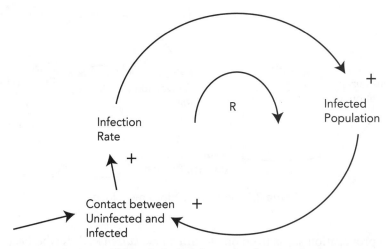

Figure 8.4 Reinforcement loop.

rate" within our company? If anything, we would like the *kaizen* beliefs and principles to propagate as quickly as common flu within an organization. What generates an exponential rate of infection is the existence of a reinforcing loop (Fig. 8.4). As more people are infected, newly infected people contact a larger number of uninfected people and infect even a larger number of people, ever increasing the infection rate. Within an organization, we want people to act as "viral agents" to spread what is good to others, creating more viral agents.

This becomes a reinforcement loop that ends only when all the population is infected. For this reason, it is important to develop future leaders who are more than skilled with the tools and techniques of process improvement and change management. We need leaders who possess a conviction and passion to develop other people and shape the culture. The practice of rapid promotion of talented managers is both an advantage and a disadvantage in this regard. On the one hand, these people move quickly across the organization acting as viral agents, yet, on the other hand, they are often promoted for results rather than for strong understanding and passion for process. A positive-reinforcement loop (Fig. 8.5) for a *kaizen* culture works in a similar fashion.

This usually starts with a few *kaizen* practitioners, perhaps with one *sensei* who teaches *kaizen* to others through *kaizen* events and projects. There may be a formal train-the-trainer program or informal mentoring.

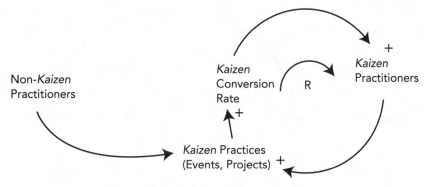

Figure 8.5 Positive-reinforcement loop.

The propagation starts from one to three new *kaizen* leaders who become "viral agents" to spread the knowledge (Fig. 8.6). As the newly trained *kaizen* practitioners practice *kaizen* together with more people, the number of people who have learned by doing increases exponentially.

A concrete example of this reinforcement loop is the practice of *kaizen* events as learning-by-doing mechanisms. If anything, *kaizen* events should be about the "infectious propagation" of *kaizen* values. Well-planned and well-executed *kaizen* events allow people to learn and practice *kaizen* behaviors and principles that shape a *kaizen* culture, not merely applying tools or copying systems from books. It is the underlying core beliefs that create sustainability. However, the opposite can happen and, unfortunately, does too often. When *kaizen* events are designed with the purpose of only improving processes or driving financial savings but not challenging people to develop their skills, *kaizen* becomes noninfectious.

Another example of a positive-reinforcement loop is the practice of daily management (Fig. 8.7). In this case, there are two interacting processes:

1. The process of making problems visible
2. The process of solving the problems

These two together drive the positive-reinforcement loop through daily management.

For instance, when a team member can't meet the standard, he or she is obligated to *stop and call for help*, perhaps by turning on the *andon* light. This process makes the problem visible. The problem is defined as not being able to meet the standard or not meeting the target result. For this process to

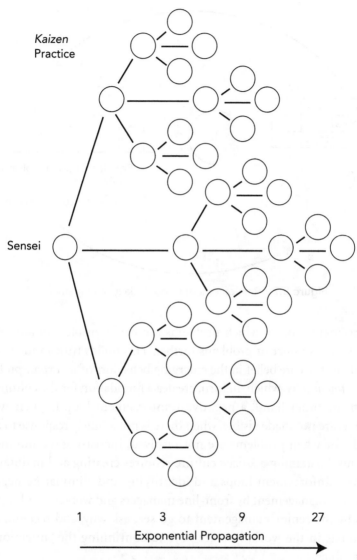

Figure 8.6 Propagation through viral agents.

work, the team member must know the actual performance *and* the expected performance and be able to compare both. The team leader or supervisor arrives to assess the situation. With the team member, the leader reviews the problem and follows the standard problem solving process to identify root causes. In this way, the team member and the leader interact in a positive-

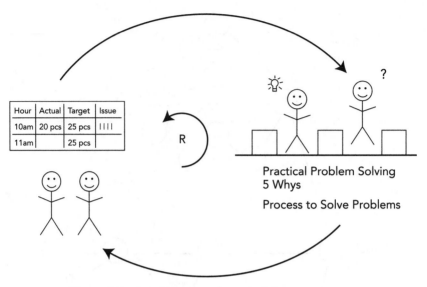

Figure 8.7   Reinforcement through daily management.

reinforcement loop in which problems are made visible and are solved, reducing the number of problems further. This builds trust in the system, strengthens the core belief in the emerging behaviors of *we expose problems* and *we stop and fix*, which, in turn, creates a foundation for the culture.

On the other hand, a negative-reinforcement loop happens when problems are not made visible, when there is not a timely response to a call for help, or when problems are not resolved, increasing the number of problems. Sustaining a *kaizen* culture requires creating and maintaining positive-reinforcement loops and identifying and eliminating negative ones. Daily management by front-line managers and workers, and support *kaizen* by the senior management to go see, ask why, and recognize the individuals in the workplace are keys to continuing the "infection" of *kaizen* values.

Haidt (2006), in writing about the *happiness hypothesis*, provided a very useful analogy about how our emotional and rational sides interact in sustaining change. Haidt depicted our emotional experience as an elephant that overpowers its rider, our rational mind, anytime there is disagreement (Fig. 8.8). Heath and Heath (2010) elaborated on the analogy and stated that "changes often fail because the Rider can't keep the Elephant on the road long enough to reach the destination." On a personal level, most of us

Rider: logical, conscious, verbal, thinking brain

Elephant: emotional, unconscious, visceral, emotional brain

Figure 8.8 Elephant and rider.

have tried and failed to maintain positive changes such as quitting smoking or engaging in proper exercise regardless of how much we rationally supported the change. When our powerful emotional elephants did not agree, we failed to change our habits. This is also true at an organizational level. This is why a "burning platform," or a situation that creates both rational and emotional urgency can work well to drive a transformation. Our survival instinct is a strong force that, when channeled properly, makes us change when the only other option is certain extinction.

As described in Chapter 7, to maintain a *kaizen* culture, leaders have to engage people at both emotional and rational levels. Without motivation of our emotions, *kaizen* culture can't be sustained, no matter how much rationale we provide. Great leaders know how to do this. Taiichi Ohno was a master of managing the emotional side. Ohno (2012) revealed the purpose behind his tactic of scolding supervisors in front of their teams as a way to drive workers to support their leaders: "I was told that you must not scold supervisors in front of the workers, but I was doing the opposite on purpose. . . . when the workers see their boss being scolded and they think it is because they are not doing something right, then the next time the supervisor corrects them, they will listen." We are not proposing using the exact same tactics, rather to engage people at the emotional level to jockey the elephant within all of us.

Modeling the Desired Behavior

To sustain a *kaizen* culture, we must make the desired behavior clear and explicit. For instance, the *kaizen* core belief in leading with humility is a very powerful one, but it does not by itself provide a clear description of the expected behavior. It is subject to too many interpretations. What does *humility* mean exactly? How do we behave in a meeting or talking with others? As leaders, what exactly should we do? Certainly the opposite is arrogance, but "don't be arrogant" is not enough of an answer.

Instead of asking others to lead with humility, begin by modeling the desired behavior. For instance, the *kaizen* habit to "Ask why five times, and carefully listen without telling them what they should do" can be an example of humble, open-minded inquiry. Asking the five whys instead of telling the answer is more than an example of humility. How often do we ask, "Who is responsible for this problem?" so that we can assign blame? How often do others quickly tell us that some other person was the cause of the problem? When this happens, what is our response or action? Do we simply assign blame to that person, or do we make an effort to understand the process that caused the problem and show this belief in our words and actions?

Another example may be not trying to jump to world-class levels too quickly, unless your leadership culture is strong enough to bounce back from failures. Each organization must search these questions, test the answers, and adapt the appropriate ones.

After three decades of working with hundreds of companies on their improvement journey, we see a common denominator that affects the sustainability of a *kaizen* culture. This common denominator is how we spend time as leaders. Simply ask yourself, "How much time each day do I spend on activities working 'in the business' and how much time do I spend working 'on the business'?" The same questions can be applied to every employee within the organization. How much time are we meeting and maintaining basic standards, and how much time are we improving them? Too many of us spend the majority of time focused on delivering the service or getting the product out the door. Some days it seems that all our effort is spent on corrective activities such as finding replacement material, getting machines back up and running, recovering defective services, and the list goes on and on. Even in our meetings it is typical that most of the discussion is centered on metrics related to output and we give little attention to process metrics and the *kaizen* activities that affect them. As we have pointed out already, notably with the Doom cycle, fixing problems without setting a new standard that is monitored and maintained is not *kaizen*.

To sustain *kaizen*, we must make time for *kaizen*. This may require that we apply some 5S—or prioritization—to our day and sort out activities that are not really needed in order to create available capacity to add *kaizen* time. It also may require that we examine our improvement activities and honestly evaluate whether we are truly working on *kaizen* or simply fixing things, caught in the Doom cycle. Although it seems like a simple action to make time for *kaizen*, in reality, this is a difficult task for many of us. Pressure to get results now can hamper our best intentions. It seems that we must focus all our efforts on getting the work done without stopping to see how we can improve our process. But this is like saying that we are too busy driving our car to stop and refill the tank.

Sustaining Through Communication at Vibco

Established in 1962 and based in Wyoming, Rhode Island, Vibco Vibrators manufactures industrial and construction vibrators. As CEO of Vibco, Karl Wadensten is an outspoken advocate of *kaizen* and lean manufacturing, hosting a radio show, "The Lean Nation," and giving many speeches in person and via video. At Vibco, he spends most of his time on the *gemba* with his people. After more than a decade in pursuit of continuous improvement, he has learned the importance of communication to make sure that key points are understood:

> The key to communication about *kaizen* activities is twofold— besides just spoken words, we use pictures and diagrams to help people understand the target condition and what we are looking to change. That way we get an emotional connection with everybody and get them on the same page, so there's no confusion or fragmentation. This is very important for us because we're trying to get everybody in the organization to affect change, not just a small number of people.
>
> In order to stay competitive, we took on a mandate that we called "same day, next day" [Fig. 8.9]. When I first introduced this to everybody, people were nodding their heads, but when I asked

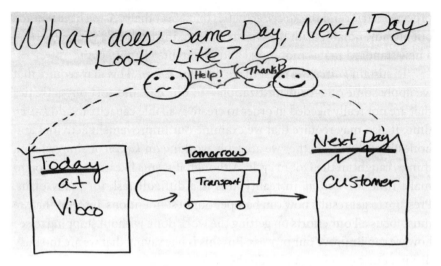

Figure 8.9 Visually communicating "same day, next day."

what this meant, I got a lot of different answers. That's when I drew a picture of the factory, a truck, and the customer site [Fig. 8.9]. Then the lights went on—people could now say, "Oh, that means we have to clear the way so we can get everything on the truck for tomorrow, and the next day, the customer gets it."

A lot of what we communicate is very tangible, physical things. This is very time-consuming—it's talking, showing what the other side looks like, actively listening and watching to make sure they get it, and then drawing pictures to help with the rest of the people. Personally, I think the best leaders spend most of their time with their people.

One of the hardest things to communicate is the idea of incremental improvement. We have to find ways to narrow the scope so that it's not so big that people don't understand how they can move the needle. Six or seven years into our journey, that message really got through, and we had something that people could really believe in.

Today, after 12 years of *kaizen*, we rejoice in seconds, minutes, single steps, saved keystrokes. When you walk out and see a machine operator, and he's talking about saving two seconds on something, and he's giddy like it's Christmas day, you just can't help but be an engaged president and leader. He's emotionally connected with the profound impact of these two seconds over thousands and thousands of pieces over the year. That's huge.

These insights are significant because many change efforts, such as the implementation of lean manufacturing, struggle when they fail to connect with people emotionally or when even the best communication stops at the level of project teams and experts responsible for process improvement. Effective communication about *kaizen* culture must be aspirational—giving people a vision, hopes, and dreams of what is possible through teamwork and continuous improvement. Based on the experience at Vibco, we can extrapolate these general rules for how to communicate in order to continually engage the hearts and minds of people and sustain a *kaizen* culture:

1. Say what is important.
2. Check for understanding of the message.
3. Check for an emotional connection to the message.
4. Adjust how the message is delivered if there is a gap.
5. Fine-tune and focus the message even on the smallest details over time.

Needless to say, communication of this type cannot be delegated from the CEO to management of an organization. There must be an emotional and rational connection from the top of the organization that motivates communication, understanding, and engagement in pursuing excellence. Communication is not about what we say but is about what the intended listener hears. This is best expressed in the words of former South African president and Nobel Prize–winning peace activist Nelson Mandela: "If you talk to a man in a language he understands, that goes to his head. If you talk to him in his language, that goes to his heart."

Sharing to Sustain

Kaizen ideas are copied across Toyota plants worldwide through a process called *yokoten*. This simply means "applying horizontally" or "peer-to-peer sharing." As part of the act/adjust phase of the plan-do-check-act/adjust (PDCA) cycle, there is the question from management, "Did you do *yokoten*?" Did you look for areas with similar problems where we can share what we learned here? This is a built-in part of the problem solving cycle and a standard practice of every *kaizen*, successful or otherwise.

Toyota has shared openly with the world since the 1980s what is good about the company, often to the puzzlement of competitors, who wonder why the company would do this. Toyota has sent employees to help other companies with *kaizen* through its Toyota Supplier Support Center, as we will see in the Herman Miller example to follow. Employees have also volunteered to help design the Nagoya Airport on time and under budget, and to the public sector through the University of Toyota. Many of the company's factories offer tours to the public and even one-day educational seminars on the Toyota Production System. By taking *kaizen* outside the four walls of our organizations, we will undoubtedly deepen our understanding of and commitment to *kaizen* and its core beliefs and increase our chances to sustain. These core beliefs include a broad respect for humanity and a desire to develop human potential, service to customers and society, urgency and dissatisfaction with broken processes and systems (which are everywhere), continuously connecting processes and people, taking action after developing consensus with various stakeholders, and sharing what is good about our organizations.

This happens sooner than later to the degree that customers and suppliers become involved whenever they have a stake in the successful outcome of an end-to-end process redesign. We encourage benchmarking exchanges and best-practices sharing because the benefits go both ways—to the visitor and to the visited. There is motivation and pride in showing off an excellent workplace. But the sense of accomplishment and satisfaction when people from other parts of the company and even outside the company ask to visit your workplace to see the examples of *kaizen* is only the most superficial reward of sharing. The importance of spreading the *kaizen* culture beyond the organization's walls goes far beyond the recognition and motivation that this provides for the organization. From the point of view of sustaining, the sharing of best practices creates a motivation to always be tour-ready because one never knows when a group of guests will come to see and learn.

The act of extending continuous improvement into the surrounding community represents an entirely higher level of commitment by leadership—committing personally, committing to mobilize the entire organization, committing to customers and suppliers, and committing to those with whom we share society but have no immediate and obvious profit-and-loss relationship. Once we begin volunteering our time to teach others about how to improve their skills, their processes, and their lives, it is no longer just a company matter; others in the community are relying on us for their personal and professional transformations. This inspires us by calling on our higher selves in pursuit of a higher long-term purpose. It is just this type of sharing and outreach activity that resulted in Herman Miller starting on its journey of transformation.

Top-Down Leadership Support at Herman Miller

Headquartered in Zeeland, Michigan, Herman Miller is a publicly traded manufacturer of office furniture, equipment, and home furnishings. Established in 1905, it had grown to a large international corporation that was struggling with growing complexity in its supply chain and delivery processes. In 1996, Herman Miller began an initiative to reinvent its operations. One early step was to visit a supplier who was also a Toyota supplier, followed by implementation of the Toyota Production System under that supplier's tutelage. This led to a relationship between Herman Miller and the Toyota Supplier Support Center.

Through a decade of the pursuit of operational excellence, Herman Miller has been able to reduce manufacturing square footage, product lead times, and inventories while growing sales and profitability. Average standard product lead times have been cut from eight to four weeks. In 2012, *IndustryWeek* named Herman Miller among the top 50 manufacturing companies in America. The Herman Miller Production System (HMPS) was extended beyond operations and renamed the Herman Miller Performance System in 2009. Ken Goodson, executive vice president for operations, initiated the highly successful operational excellence program at Herman Miller and now heads the company's operations worldwide. He shares his insight into what is required for continuous improvement to succeed long term:

> There's one thing that will determine whether any continuous-improvement program succeeds over the long term, and that is top-down leadership support. Period, end of story.
>
> Everybody here knows that if you work in operations, you will use these techniques, or you will not work in operations at Herman

Miller. Now that's a pretty strong statement. People might ask, "Is this just a dictatorial process to control people and get them to work harder?" and the answer is "No."

First of all, dictatorial processes work for a very short time before people rebel. Second, Herman Miller, since the 1950s, has been a participative environment, meaning that people have a right to influence their workspace. People have the right to demand answers to questions as to why they're doing x, y, or z.

The continuous-improvement process gives workers the tools to help change their workplace for the better. It allows them to become engaged in improving safety, quality, and how they do their work. But without top-down support—strong leadership support around process—this will simply fail because it is hard work, and hard work doesn't sustain itself. We become lazy; we go back to the old way. Even when the results are fantastic, it's still hard to do. It's hard to be an Olympic athlete and stay an Olympic athlete. We've been able to sustain the effort through top-down support.

The ability of an Olympic athlete to remain at the highest level of competition is limited by his or her physical and mental fitness, which for most Olympic sports is the result of two decades of work at the very least. Creating and sustaining a *kaizen* culture over many more decades are both easier and harder than the challenge for individual Olympic athletes. Shaping an adaptive organizational culture requires that everyone in the organization, not only the star athletes, participate in daily practice, identify with a common purpose and work together as a team, and continually develop the next generation of world-class performers. This requires the continuous development of coaches within the organization.

The Trick to Sustaining Kaizen at Kaas Tailored

Kaas Tailored is a manufacturer of furniture for restaurants and hospitality, as well as aircraft interiors. Employing 150 people from 14 countries, Kaas Tailored is one of a decreasing number of manufacturers in the United States that does high-quality cut, sew, upholstery, and custom furniture manufacture. Established in 1974, Kaas Tailored has been led by second-generation President Jeff Kaas since 1997. Since 2000, Kaas Tailored has been on a journey to go "One Step Beyond" for customers and employees. *Kaizen* has been a major part of the development of Kaas Tailored.

Jeff Kaas encountered *kaizen* when one of his major customers, aerospace giant Boeing, contacted the company to offer to help in reducing costs. Boeing offered to send internal consultants to conduct rapid process-improvement (RPI) projects to help Kaas meet cost-reduction targets set by the purchasing department at Boeing. The RPI project was a direct descendant of the five-day *kaizen* event that the Boeing Company had learned from Japanese consultants from the Shingijutsu Group. The offer was to find opportunities to reduce costs by examining the processes at Kaas Tailored to remove waste rather than taking the costs out of the company's profit margin. Seeing the advantages of understanding the company's processes and applying *kaizen* before inviting Boeing in to do the same, Jeff Kaas contacted local *kaizen* experts.

The *kaizen* learning process started by teaching the management team about *kaizen*, waste, and problem solving through an eight-week course, meeting during the lunch hour so as to avoid taking supervisors and managers away from their jobs for too long. The two *kaizen* consultants challenged Kaas Tailored to look at its lead time from receipt of an order until shipment of the order for custom furniture and reduce it from eight weeks down to three weeks. The walls of the meeting room were covered with butcher paper, and the Kaas Tailored management team was encouraged to lay out the process step by step, color-coding activities that added value to the customer (green), those that added cost but no value (red), and necessary non-value-adding steps (yellow). From this, a prioritized action plan was created, and *kaizen* began.

Thirteen years later, Kaas has maintained constancy in its long-term purpose. The statement of purpose is slightly updated yearly, and in 2013 it has three elements:

▲ 20 by '20.
▲ Know and show the truth.
▲ Improve every day.

The first is a growth goal to $20 million by the year 2020, a results metric; the second is a process metric; and the third is a mix of both. As we saw in Chapter 5, the annual breakthrough objectives for the "improve every day" long-term policy included "cutting *muri* by 50 percent, tripling the value of our routines, and true pull." *Muri* is the Japanese word for "overburden" or unreasonable work conditions. Specific projects to support these are generated bottom-up by the people who understand the processes best. How has Kaas put the long-term statement of purpose into practice for over a decade?

Know and Show the Truth

Jeff Kaas has been to Japan five times since the year 2000 so that he could see the truth. "Truth sets you free from slavery, from sin, from waste," says Kaas. On each trip he took some of his managers with him. Returning from one study mission to Japan, Kass finally saw the walls in the office. The front office of Kaas Tailored was typical, with the functions of sales, marketing, engineering, planning, and quality having separate offices around the outside of the building, near the windows. "We just had to trust the principles enough to say 'It's not flexible enough; we are always working around the walls.' The realization was that we were fretting about it, spending time studying the changes before rather than trying it and studying it afterwards."

Another major milestone was moving the toilets from the center of the factory on the second floor to the corner. One side of the factory couldn't see what was going on with the people on the other side. There was no clear financial justification for the construction expense. However, it was another step toward an organization where it was always possible to show the truth. "We didn't know what we weren't seeing until we moved the toilets," says Kaas. Visibility, communication, and speed of problem solving in the second-floor operations continued to improve (Fig. 8.10). This was also the

Figure 8.10 Showing the truth at Kaas Tailored.

change that allowed Kaas Tailored to move the desks of the office staff onto the shop floor.

There was still the problem of lack of visibility between the first and second floors. How much of a problem was this? The countermeasure was to place closed-circuit cameras in the wood shop focused on the location of finished frames. A large flat-screen monitor was placed on the wall where the toilets used to be on the first floor, linking the upstream and downstream processes with the judicious use of technology.

Improve Every Day

There are 12 languages spoken at Kaas Tailored. This is not just atypical, but remarkable, for a manufacturing company in the United States employing 150 people. Sewing and upholstery are skills in decline in the United States but in much demand at Kaas Tailored. Many talented sewers and upholsterers come from the local communities of immigrants from various countries where these industries and skills are vital. Not everyone speaks English fluently, yet at Kaas Tailored, everyone is engaged daily in the maintenance and improvement of standards. All team leaders are responsible for helping every team member put in place one *kaizen* idea per month. These *kaizens* can be improvements people made not at work but at home (Fig. 8.11). "We

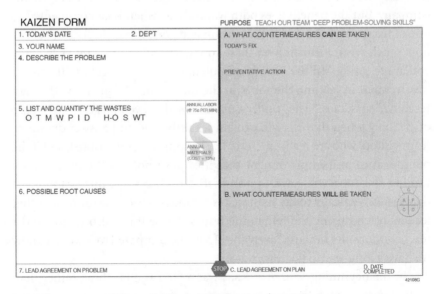

Figure 8.11 Improve every day at Kaas.

are trying to grow people and their problem solving skills and not just get results at work," says Kaas. During team training, when peers rotate responsibilities to reinforce a standard by training each other, if a person cannot speak English, he or she will have an interpreter. There is no excuse.

"The power is in having routines," says Kaas. "We just do it, and ask questions later. For example, cleaning together as a team for five minutes each morning, they [workers] may not realize they are eliminating waste and maintaining standards, but they are. By doing pull, they are keeping waste out—they are contributing. There is a saying in the Bible, 'As a dog returns to its vomit, so fools repeat their folly.' People may set up something perfectly for themselves, do it for a while, and think it's the best thing ever, but then go back into their old habits. The routines are the daily bread, the reminders of what we need to do."

Sharing What Is Good

When asked why he opens his company to tour groups three days every week without compensation for the time, Jeff Kaas replies, "Love your neighbor." Sharing what is good at Kaas is an activity deeply rooted in values. More than 2,000 people per year from the community and across the country visit Kaas, and the calendar is booked six months in advance. Jeff Kaas teaches others and shows them how *kaizen* has changed his company. But the benefit goes both ways. Once, when Kaas commented to one of his supervisors after a tour group had gone through, "I can't believe how much we are learning on these tours," the supervisor replied, "Only you are learning." In the next *hoshin* plan, Kaas set a target of 50 percent involvement in leading the tours in the next year. "People grow through teaching," says Kaas. "They understand *kaizen*, the seven wastes, the purpose of cleaning when they have to explain what they do and answer questions from visitors who are really interested." The tours are not scripted, and if the maintenance and improvement activities have not been kept up, those leading the tour will have to admit that their area is dirty, sloppy, or dangerous. Typically today Jeff teaches for one hour of the tour, and other leaders teach during the remaining hours. These tours also reinforce the teachers' humility because "even after 13 years, compared to Toyota, I know we have a long way to go," says Kaas.

In the beginning, *kaizen* at Kaas was a "Jeff wants" thing rather than clear alignment with the long-term purpose of the company. The company saw and copied what it understood from Toyota, and learning happened afterwards. Today Kaas is no longer copying what we would see at Toyota; it has developed its own unique culture and management system to grow its people and its business. "You need to be accelerating your growth to sustain—companies are either growing or shrinking—there is no staying in place. All the *kaizen* tools and routines are designed around growth and challenge, like coming up with one idea per month. There is no trick to sustaining," says Kaas. "Just always improve."

The Way We Do Anything

As we have noted, what we call "lean" today is an adaptation of the Toyota Production System. The Toyota Production System was built through the relentless practice of *kaizen* across an automotive supply chain and its management practices. The practice of *kaizen* is an emergent behavior of *kaizen* core beliefs. These *kaizen* beliefs, we propose, derive from the human motivation to be good and do well. This takes no time at all to say but a lifetime or more to realize. The shaping and sustaining of an adaptive culture of excellence must be approached with an understanding of the whole and the details—that the way we do anything is the way we do everything. The simplicity of the PDCA cycle allows its adaptations to all levels of action, from the long-term purpose to the annual plans, from quarterly performance to improvement projects, and from daily *kaizens* to personal habits. When this is clear, it is like a fire is lit inside. In the words of Taiichi Ohno, "Understanding means doing." In order to build an enterprise of sustainable excellence, we must improve our odds of survival through proper planning and preparation of the organization and its people, the subject of our next chapter.

CHAPTER 9

Organizational Readiness for *Kaizen* Transformation

If I had eight hours to chop down a tree, I'd spend the first seven sharpening the axe.

—ABRAHAM LINCOLN

If we waited until we were completely ready to make major changes, many opportunities would pass us by. We can't stop and wait to make improvements until everything is ready. But whenever we decide to start on a transformation—that is, on a major change that affects many people, is highly visible, and for which there are great expectations—it is important to prepare thoroughly. Whereas, along the way during a transformation, small failures are acceptable, when a transformation fails, that is, when an organization stops believing and gives up, not only have time and energy been wasted but negative beliefs and assumptions such as "we just go from one program of the month to another" to "leadership isn't serious about change" to "my ideas don't matter," have also become part of the culture. In order to be in alignment with the *kaizen* principle of planning slowly and thoroughly and acting with urgency and certainty, we must take the time to ensure that certain basic organizational readiness preparations have been made so that we can avoid the major pitfalls and give ourselves the best chance for success.

Whether we are just starting or reflecting on the progress of our continuous-improvement journey to excellence, it is worthwhile to reflect again on the culture of the organization—how we make decisions, how we involve and connect people, how we improve, how we follow through, and so forth—in order to design experiments that demonstrate early success in changing not only physical or technological elements but also bad habits.

Although there are many successful approaches to change management, we have arranged organizational readiness activities for accelerated change under five themes:

1. Top-management commitment
2. Alignment of *kaizen* (or transformation by whatever name) with long-term purpose
3. Stability, safety, and security
4. Middle-management readiness
5. Making a smooth start

Each leadership team will need to face reality with open-minded humility, learn from others, and use their best judgment to decide how much time and energy to spend on the details of each theme.

Top-Management Commitment

A simple rule of thumb that applies to organizations of any type or size is that the top 50 leaders and people influential in the organization must have strong belief in the transformation. If the company has only 50 people, this would mean everyone, because if even 10 of 50 people failed to pull in the same direction, this would be a big drag on progress at this company. For a 500-person organization, the senior leadership team, middle management, and informal opinion leaders all must be engaged. The definition of *top management* is fuzzy in this regard—we are not talking about only the CEO, his or her direct reports, and senior vice presidents, but enough levels down into the organization to make the information gathering and feedback real and practical. Any transformation must be strongly supported and guided from the top, and by having the top team extend to a number as large as 50, this allows for a pilot team to practice behavior change not as an isolated project but within the natural course of business.

However, this initial learning, awareness, and commitment phase must happen rapidly in order to maintain interest, momentum, and belief that change will actually happen, ideally within three months from signaling a genuine intention to start. It is better to have a longer period of study and learning by trial and error before a formal commit to a large program is made so that a shared top-management commitment experience can be designed and executed. During this study and trial phase, *kaizen* activity at various levels will deliver many of the desired business results. However, this may not be enough to create the strong belief among the top management to bring about commitment to behavior changes that sustainability requires.

This study and trial phase experience must achieve several things for the leaders. First, it must allow them to engage personally in the change and learn what it is they will be leading and championing in the future. Top managers must be able to articulate clearly how the transformation fits the business strategy and develop a passion for and understanding of what they need others in the organization to do before even considering delegating responsibility for execution. They must commit the time and energy to study, and while this can be very expensive time, the cost of skipping the investment in top managers will be far greater, paid in the form of transformation efforts that stagnate, do not achieve their potential, or are abandoned.

Second, the leaders must understand the affected processes in their current state and how the transformation will affect them. Improved business results come from improved processes, and improved business processes come from people who understand them and how to apply the scientific method to make improvements and innovations. Leaders must understand *kaizen* and how it applies to the organization's current situation, and this is best learned hands-on.

Third, having a broader group of leaders—the top 50—going through the awareness and learning phase actively and visibly, makes the effort more credible and practical when it comes to engaging all employees, top to bottom, in shaping the desired culture through concrete action. In contrast, when only the very senior executives go through workshops, off-sites, and strategy sessions—often with inadequate go-see fact gathering because of their busy schedules—the outputs can seem disconnected to the next levels of management. Also, not engaging multiple levels in this learning limits the opportunity for *hoshin kanri*–style (strategy deployment) two-way communication and catch ball as part of the process to link high-level objectives with concrete actions that are needed at the front lines.

As part of this learning and awareness phase at the very beginning or even before officially kicking off a transformation effort, we recommend a "look in the mirror" process to come to grips with the reality of the readiness of the organization's change capability. This can be through storytelling and reflection on successes and challenges with similar large-scale efforts in the past, such as a simple 16-point three-level scale of effectiveness (Fig. 9.1).

No	Change Capability Questions	Score
1	We start with a clear picture or vision of the future whenever we make a change.	
2	We allocate appropriate and sufficient resources needed to make the change successful.	
3	We actively involve people who are impacted by the change in designing the future state.	
4	We involve those people who are most affected by the change in identifying potential obstacles.	
5	We take steps to ensure that people affected by a change have the knowledge, skills, and emotional states necessary to succeed.	
6	We establish processes to document, monitor, and report our progress in the change to stakeholders.	
7	We clearly communicate the purpose and rationale for the changes to everybody.	
8	We communicate new standards, norms, and expected behaviors during the process of change.	
9	We confirm that people have a clear understanding of the standards and expectations that accompany a change.	
10	Throughout the change process we regularly inform people how well they are meeting the expectations of the change.	
11	We establish channels for ongoing two-way communication and feedback between leaders and those who are led.	
12	Leaders make themselves easily accessible to listen, answer questions, and share information during the process of change.	
13	We create opportunities for people to test, practice, and rehearse the new behaviors or actions through pilots, simulations, role playing, or visualization.	
14	We pilot, test, or trial changes that are significant and far-reaching in order to learn and make adjustments, before rolling them out across the organization.	
15	We acknowledge, recognize, and celebrate the efforts, milestones, and achievements during the change process.	
16	The organization has more energy as a result of implementing the change.	
	Total score (out of 32)	0

2 = very true
1 = true
0 = not true

Figure 9.1 The organization's change capability.

The purpose of this exercise is neither to beat ourselves up nor to tell ourselves how great we are. We are exploring a transformation effort presumably because there is a business need, and to give ourselves the best chance of success, we need to learn about ourselves and how we are likely to manage this change. One useful reference point is to see how we have handled similar efforts in the past. A final word of advice to top-management teams engaged in the learning, awareness, and commitment phase: if it doesn't feel like work, you probably aren't getting it yet, and neither will the people you lead. A transformation should stretch people, and if the team isn't finding the new way of working difficult to adapt to, chances are that the change is still superficial and not at the level of behaviors and cultural assumptions. The faster that people are engaged in the process, the higher is the probability that behaviors will change and the culture will shift. A top-management team that is engaged and able to talk about what it means for each member personally will be able to talk in plain and practical terms without losing people in jargon developed 30,000 feet up in the organization.

Aligning *Kaizen* with Long-Term Purpose

One of the responsibilities of leaders guiding an organization through a *kaizen* transformation is to give clear direction and communication on how continuous improvement fits in with the purpose and plan of the business in the long term. Periodically, the organization will need to renew the clear purpose or add detail based on new understanding or new horizons. Many organizations adapt *kaizen* as one of many improvement tools or a format for making rapid changes within operations. While this is not wrong, the *kaizen* approach used mainly to drive business results with weaker emphasis on developing leadership skills is doomed to lose momentum and falter. Likewise, improvement activity focused mainly on the development of people gradually becomes a series of briefings and presentations without focus or alignment with urgent priorities of the organization. When either of these things happen, people at all levels lose passion for continuous improvement and with it the energy put in follow-up efforts, leading to gradual decline and the belief that "*kaizen* does not work here."

It is instructive to look back on the arc of development, decline, and resurgence of quality control (QC) circle activities in Japan. In the 1960s, small-group activities were introduced to reduce quality defects through the program called total quality control (TQC). During the 1970s, the efforts gained momentum, with the Japan Union of Scientists and Engineers (JUSE) codifying the philosophies and approaches of QC circle activities. This was very important in laying a foundation of quality that supported the development not only of the Toyota Production System but also of Japanese industry in general. No small part of this success was due to the three fundamental principles of QC circle activities articulated by JUSE. As their long-term purposes, QC circle activities must

1. Contribute to developing and improving the health and resilience of companies
2. Respect humanity and create cheerful, lively workplaces
3. Bring out the unlimited capabilities and possibilities within people

These beautiful guiding principles influenced the development of *kaizen* cultures across hundreds of companies and engaged tens of thousands of people across Japan in all industries. However, in the 1990s, with the bursting of Japan's economic bubble, industry was faced with excess production capacity and excessive debt, and many were forced to do the unthinkable and restructure their labor force. The result was a near doubling of part-time and contract labor to a third of the Japanese workforce over a period of 20 years. The impact of this change was to sap the energy from QC circle activities because of the challenge of involving these part-time employees in these activities.

The bursting of the economic bubble and the so-called lost decade of slow growth in Japan left corporate leaders at a loss as to how to refocus the purpose of their QC circle efforts. Most companies spoke about the total quality management (TQM) building blocks of customer first and never-ending improvement, but in practice they balked at investing in total involvement. Toyota, in contrast, doubled down. Faced with the challenges of an increasingly global business, in 1993 the company introduced new QC circle activities with stronger emphasis on each individual playing a role, and in 1994 it launched a TQM training course for new executives within the Toyota Group, followed by an announcement of the adoption of TQM at the All-Toyota Total Quality Management Convention later in the year. Toyota's deep historical and cultural commitment to respect for the individual and the development of human capability no doubt allowed the company to continue investing in QC circle activity rather than simply focusing on technical *kaizens* that were more results-driven.

Ironically, QC circle and *kaizen* activities saw an increase in the United States and Europe in the 1990s and beyond under the names of *TQM* and *lean*, which continues to expand slowly across the world today. It is too early to tell whether organizations around the world that are practicing TQM, six sigma, lean, and other popular variations of business excellence are doing so out of a clear and renewed purpose to serve customers, strengthen the company, improve the work environment, and develop people, or whether these are still seen largely as mere tools, another means to increase short-term profits. Organizations that fail to strongly align the long-term purpose, as seen in the fundamental principles of TQC articulated by JUSE, deeply within their values and cultural assumptions are at risk not only of failing to draw out the maximum potential of their people but also of losing their way during times of economic crises, as even some of the best Japanese firms have done.

In practical terms, what time horizon should the modern organization use for "long-term" thinking? This is perhaps the most challenging issue in regard to the successful design, deployment, and sustaining of business excellence today. Too few large corporations have any long-term constancy of purpose, in part because of the strong pull from the investor community for financial returns on a 90-day cycle. A socioeconomic transformation in that area is beyond the scope of this book, but what we can advise is that the leadership must be committed to seeing the transformation through until it becomes part of the culture and is not dependent on a small group of influential leaders to keep it going. Once people have formed new and better habits, have experienced success as a result, and have built strong belief in the new ways, the artifacts, behaviors, and assumptions have been integrated.

How long does this take? In *kaizen*, results come quickly, with significant performance improvement seen in a discrete area within a week and across entire operations within 6 to 12 months of dedicated effort. Making these results sustainable is what requires developing people, and this can take an additional two to five years to go from basic competence through practice to mastery and integration into the standard, nonnegotiable leadership practices that are taught from generation to generation. However, even before starting the clock, stabilization and organizational readiness activities can require six months to a year depending on specific issues for the organization. These activities to stabilize business processes and financial or human resource issues are also improvement activities, so this does not mean that there is no progress during the readiness phase, only that structured, accelerated improvement should wait until one is ready. At first we crawl. Then we learn to walk before we learn to run. And when an infant begins to crawl or walk, we must baby-proof the house to avoid injury from the inevitable falls. Leaders must make an honest assessment of the change capability of their organization when setting both medium and long-term goals for *kaizen* transformation.

Believing in Service Excellence at the Rotorua District Council

The city of Rotorua, located in the central North Island, is the mountain-biking capital of New Zealand. Created in 1979, the Rotorua District Council (RDC) is the local government body of the city and county. Employing a staff of 550 people, the RDC is responsible for providing a wide variety of services to the public, including animal control, building services, emergency management, economic development, environment and health, elderly housing, business licensing and permitting, planning and control, pollution control, taxation, recycling, roads, waste management, and water. The RDC launched a lean transformation initiative in 2011 in response to a government statute that required for the first time that local governments improve quality, efficiency, and effectiveness. "We had to improve performance in order to meet expectations of our community," says Peter Guerin, chief executive.

The RDC was not new to continuous improvement. However, "None of these systems talked to each other or to the management team. We followed the private sector patterns in adopting various business-improvement methods beginning in the 1990s," recalls Mijo Katavic, business improvement and innovation manager. "This included TQM, and later ISO for our labs."

"I was convinced that we needed to have a corporate approach to continuous business improvement," said Guerin. "We studied several methods, and the lean approach was the one that stood out." Although he wanted to begin implementation immediately, Guerin took the time to condition the organization for what was coming. Communication happened through a series of weekly newsletters with examples and quotes about lean on the front page, a speaker from Waikato University, a two-day workshop with the extended management team, and discussions by Guerin with groups of 20 to help the staff understand what was being proposed to get their feedback. Guerin also engaged the organization in an exercise to streamline the number of departments, from four to three, in preparation for a more customer-focused process. "As a public-sector organization, we used to be very resistant to change," says Guerin. "For example, a few years ago we had 13 reception areas across multiple floors that handled 20 percent of the customer traffic and—we used to think it was good customer service. Now we have one customer center that deals with 80 percent of the customer traffic. Results like these have helped us to embrace change. I had to temper my urgency and eagerness to make changes in order to build a platform of readiness. If I had to do it over again, I would have gotten into implementation quicker."

As a typical government organization, the RDC was very silo-based. Senior experts would be at the head of a functional department with technical people for that function, rarely working with other functional areas. "We now have what we call our 'Hall of Councils,'" says Guerin. "The aim is to have a 'one-stop service'—the customer should not have to deal with multiple departments. If the one stop is not able to provide the answer, that person is responsible to escalate the issue to their colleagues in the RDC in order to meet the customer's need." Silos continue to be a challenge, but one of the points of positive feedback from the staff has been that the interdepartmental work is increasing. "This is something that never happened before we put the focus on serving the customer better as a whole organization," observes Katavic.

Having different functions working together also has helped the staff to see things from the customer's viewpoint. Bringing the economic activities group, which includes the museum, tourism, marketing, and economic development, to the regulatory side, such as building, planning, and licensing, was an eye opener (Fig. 9.2). "They began to understand what delays cost the customer," says Guerin. "In the past, the focus would have been on quality, on taking time. Now they are getting the balance of quality and cost from the customer's viewpoint."

Although early on its journey of adopting lean practices, the RDC is conscious of involving the wider community. "When we have a *kaizen* event, we try to get a third of the people from the process, a third from manage-

Figure 9.2 Cross-functional *kaizen* event at the Rotorua District Council.

ment, and a third from customers. For example, during a recent *kaizen* event in the land-development process, we got a builder of garages who helped to redesign the licensing process," said Katavic. The RDC raises awareness of improvement savings in water and energy through visual management in public spaces and has begun promoting *kaizen* to businesses within the district so that they can improve their performance.

The RDC continues to work on strengthening the foundation after 18 months, with more than 50 percent of the staff having experienced either formal lean training or a *kaizen* event. Everyone has been informed via large-group face-to-face communication and monthly electronic communication. New staff members are given orientation to lean and a tour. "We have 80 staff certified at the *kaizen* practitioner level, and they are able to lead improvement projects within their own areas," says Katavic. There is a constant and consistent message from the chief executive about lean, "It's the way we work," and this is backed up in the biannual performance-review process. "Each person has responsibilities referring to lean in his or her performance goals," says Guerin. Fully 40 percent of the performance review is linked to how well people live the shared values of the RDC, which are

▲ *Service*. We strive for excellence, continuous improvement, and value for our customers.
▲ *Teamwork*. We communicate and cooperate with each other and with our communities—"*tatou tatou*."

▲ *Integrity.* We act with commitment, honesty, openness, and respect for others.

▲ *Recognition.* We appreciate and acknowledge effort, and we celebrate success.

▲ *Satisfaction.* We take pride in our work, and we have fun together.

"Our vision is to be an outstanding organization, and our mission is shaping a great future together through inspired leadership and superb service. Three years before we initiated lean thinking, we began reflecting on why we exist and how to provide outstanding service. Lean pushed the right buttons in terms of helping us live our values and fulfill our mission and vision," remembers Guerin.

A lesson learned was "Good process, good results." Lean thinking and what it meant for costs and savings were included in the last long-term plan communicated to the group. "Now we understand that we got it backwards. It is a fanatical focus on the customer, stripping out waste, and then the cost savings will follow," notes Guerin. One of the biggest benefits has been on the sense of confidence and engagement gained through *kaizen* activity. "We want to be the best council in New Zealand," says Guerin. "Three or four years ago we would have never said that. We did not have the tools, the ethos, or the values in place. Now we can say it. And people believe that we can get there. Some staff members think that we are already there in some places."

Stability, Safety, and Security

How do we know that this is the right time to start down an accelerated path or transformation or that we should continue readiness activities? If those asking have progressed to the step of asking this question, the good news is that the top-management team is committed, and the transformation is aligned with the organization's long-term purpose. The bad news is that before we can get started with the fun of making rapid improvements to our processes, we must take a hard look at any unintended consequences our improvements may bring. By nature, the changes to processes that come as a result of embracing the core beliefs of *kaizen* will expose problems that were hidden or turn situations that were okay into problems. There is a saying at Toyota, "'No problem' is a problem," and this means that when we can't see our problems, this is a problem because they are still there, and we are not aware of them. Many problems and issues will surface as a result of connecting processes and people in a flow, synchronizing work with internal and external customers based on a pull, and sowing a sense of dissatisfaction in the status quo among the workforce. Luckily, many, if not most, of these are technical problems with relatively simple and emotion-free counter-measures. The changes that bring people-related problems to the surface are the ones that we must be aware of and prepare for during the organizational readiness phase. These include, but are not limited to, such questions as:

▲ "Is the business viable, and is employment secure?" Often continuous improvements are part of a two- or three-prong strategy to reduce cost and increase revenue growth and may be paired with restructuring. How this issue is approached will affect how the integrity of the leadership is perceived.

▲ "Will *kaizen* activity result in job cuts?" Leaders must have a no-layoff policy for improvement activity backed up by a strategy to redeploy any people who are freed from improved processes, such as bringing in new business, insourcing, reducing overtime, cross-training of workers, new-product innovation teams, community service, or a *kaizen* pool of people to work on safety, quality, delivery, and cost targets in other areas.

▲ "What happens to my overtime income that I rely on to pay bills after productivity has improved by double digits?" This is a variation on the preceding question, made more difficult by the fact that similar amounts of overtime work may not come back as productivity improves and the workload is leveled out.

▲ "Do we have basic stability in the areas of absenteeism, quality, safety, and material availability?" Although *kaizen* can be used to establish basic stability at a process level, if instability is due to deeper behavioral or cultural issues or contract issues with customers, suppliers, or labor partners, a more direct intervention may be required first to stabilize the situation and then to give the transformation effort a fair chance to succeed.

▲ "Why do we tolerate nonperformance and/or noninvolvement?" Leaders must address root causes of this behavior and make it clear that the change is not optional and that total involvement is expected.

▲ "How well do my current management skills match need in the new system?"

The most important thing is not to have answers for all these questions but to anticipate them and be honest. This shows respect for people. Leaders must have the integrity to make promises that they can keep and follow through on those promises. We must be honest about the fact that not everyone will be happy with all changes. Within any change or rules, there will be winners and losers, and we must honestly be able to say that the proposed direction is for the greater long-term good of the vast majority of the people.

It is important to connect and align all stakeholders with the original purpose and objective of the organization, or risk resistance to change as a reaction to a perceived threat to survival. Writing about the problematic behavior of decision-making groups, Hall (1976) observed: "The problem with bureaucracies is that they have to work hard and long to keep from substituting self-serving survival and growth for their original primary objective. Few succeed." During a *kaizen* transformation, middle managers often perceive that they have less to gain and more to lose as the front-line workers are empowered, processes and problems become more visible and better understood, years-old middle-management responsibilities and routines change, and they go from being bosses to coaches. Readying this group for change, helping them to grow, and seeing them through early resistance are key responsibilities of senior management.

Readying Middle Management for Culture Change at Bosch

We have seen that operational excellence is achieved and sustained only with strong top-management commitment and support paired with an empowered and engaged front-line workforce. But what about the people who are caught in between these two groups? The so-called middle management is a broad and diverse segment of any organization that must be engaged deliberately. There are unique challenges because on the surface it can appear that members of middle management have the least to gain and the most to lose as a result of a culture shift toward becoming more adaptive and improvement-focused.

The role of middle management during a *kaizen* transformation must shift from decision makers to delegators, from problem solvers to problem-solving coaches, from subject-matter experts to process and facilitation

Table 9.1 The Shifting Role of Middle Managers

From	To
Decision makers	Delegators
Problem solvers	Problem-solving coaches
Technical subject matter experts	Experts at facilitating processes
Heard giving answers and orders	Heard asking the right questions
Source of power and fear	Powerful removers of fear

experts, and from symbols of fear and power to powerful removers of fear (Table 9.1). The changes seen in people and organization are sometimes unbelievable, nothing short of transformational, and speak to the awesomeness of human potential.

Establishing middle management's pivotal role was a top priority when João Paulo Oliveira, senior vice president at Bosch Termotecnologia, first introduced *kaizen* to his organization in Portugal. Here he discusses the cultural issues he encountered:

> When we introduced *kaizen* to our middle managers, we told them that they are the most important people in the company for the success of *kaizen*—more important than the CEO. I became very aware of this when I allowed a veteran middle manager who had no passion for *kaizen* to lead one of the first pilots, and it was a huge flop. I was lucky because we had started with two pilots, and the other went very well.
>
> We used the results from that successful pilot to help our middle managers understand that we would be providing them with tools that would make their lives very different. It is tempting to say easier, but this is not true. *Kaizen* takes a lot of discipline, and that's hard, especially in our Latin culture here in Portugal, where we see rules as something you go around. It is only easier in a certain way, because you develop some tools that allow you to focus on areas other than on firefighting.
>
> Our middle managers are normally workers with limited education who progressed up the ladder of the organization, so we give them a lot of training. Much of that is about leadership.

Delegating is an area that I am very concerned about, because managers often equate it with losing power and influence.

I try as a mentor of my people to tell them that in a large corporation, we cannot personally look after every tiny little thing. Of course, we do have to attend to some specific points. But we have to delegate, we have to trust our team, and we have to manage them through a mentoring relationship. Delegating takes a lot of practice. Like sports, you have to train at it every day and overcome that urge to control everything.

Leading by example is also very important. I was on a benchmarking tour in Japan years ago, and a plant manager was explaining his operation to about twenty of us. Suddenly, he stopped his presentation and went to pick up a piece of paper that he saw on the floor. The factory was very, very clean, and this was very easy because he set the example and people followed.

We very often forget that if we're in a leadership role—whether that is middle management or lower management or top management—whatever we do is being observed and judged by our people. That includes the way we sit, the way we talk, the way we dress, the way we act.

We're not perfect machines, of course. Sometimes you might be very tired and find it difficult to sit in the correct position for three hours. But it is very important that as a leader and manager, you are aware that you are constantly being observed. That's the hard part of being a manager. But if you lead by example, your life as a manager becomes much, much easier. I am fully convinced of that.

We can take away three valuable lessons regarding the preparation of people in middle management from the experience at Bosch. First, it is important to start small, learn, and adapt. Although one of the two pilot projects failed, this showed areas where the Bosch leadership needed to work with middle management to shore up their competence and motivation. Second, the ability of middle management to delegate responsibility is critical to developing the next generation of leaders. Third, leading by example and being visible doing the right thing speaks louder than any words. These three points are true in the culture of any organization or country.

Leaders as Teachers: From Span of Control to Span of Support

One of the greatest misconceptions made by modern management is placing unquestioned virtue in flatter organizations. On the surface, a flatter organization, because it has fewer management positions, would appear to have lower costs. Part of this is understandable because technology has streamlined communication and enabled productivity gains, but there is nothing inherently better about an organization with fewer layers of management. In fact, most of the time, these guidelines on span of control and team size of a leader seem entirely arbitrary and disconnected from how the business runs.

There is a correlation between group size and the ability of a teacher to effectively develop the students. Whereas a lecturer can communicate with a virtually unlimited number of people via one-way communication, two-way communication is immediately limited by the time and attention that a large number of students require from the teacher. The same applies outside schools. For those of us aspiring to become learning organizations, we must consider the impact of team size on the team leader's new responsibility to become a teacher. Far too many supervisors and team leaders have teams that are so large that they have no time for training, improvement, planning, and the more adaptive behaviors that make teams and organizations successful. The first step is to change our thinking from *span of control* for a leader to *span of support* (Fig. 9.3).

A leader who cannot control everyone within his or her span may not feel bad because adults should not need constant supervision or to be controlled. A leader who cannot support everyone on his or her team when they need it will feel a stronger sense of failure. The span of support, or team size, for a leader must be designed based on a deep understanding of the processes that the team is responsible for, the frequency and duration of the interruptions or problems that arise, and the capability of both the leader and the support organization to resolve these issues within the hour, day, or week. Frequently this information will not be available, and the span of support must be simply a guess, a hypothesis to be tested and refined. Some leaders may be uncomfortable with this level of ambiguity. It may take persuasion and coaching for them to accept that their team size may grow or shrink as more data are gathered on the support needs of the processes

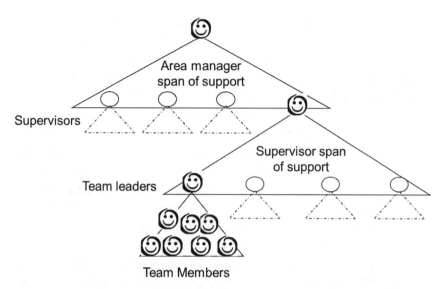

Figure 9.3   Span-of-support example.

and people; as they become better at problem solving; as processes change because of technology, customer demand, or *kaizen*; and so forth. Having a large number of direct reports no longer becomes a sign of power or privilege, only an indication that this is the number of people that leader is able to support based on the needs of the team and the leader's capabilities.

Other questions that must be asked during the readiness phase, in preparation for converting traditional managers into *kaizen*-oriented teachers, include

▲ How many management levels are there between worker and head of site?
▲ How many people report directly to the head of site?
▲ What is the span of control/support of each leader?
▲ How many organizational boundaries (silos) exist within one value stream?
▲ What is the balance of support staff between shifts?

Again, there is no single correct answer to these questions. They are aimed at stimulating thought and discussion about the environment in which we are asking leaders to become teachers. How successful can they be, and what challenges will they face? Before setting any broad policy about team size, span of control, or number of layers, leaders within a *kaizen* culture must understand customer needs and how their processes are behaving and design teams that can manage and improve them effectively. The best way to do this is to develop and test a hypothesis in a practice field or model line, where the group can test the hypothesis that a leader with a span of support that is appropriate for the process and customer behaviors can deliver higher performance than the current way.

In many organizations, as we saw in the Bosch case, middle managers rise to their position through seniority or technical capability without having received a great deal of formal training in management skills. This places middle managers at a disadvantage when the pace of change accelerates because the processes that they know how to control are suddenly changed based on the ideas of their subordinates, and the value of their knowledge and experience can seem diminished. Even when middle managers have received formal management training, if the culture of the organization does not require that everyone practice these good habits, the knowledge is not used, and it is never converted through practice into a skill.

This is the great loss of opportunity of most management training—it is a good idea but not strongly supported by more powerful routines, beliefs, and assumptions in the culture waiting for them in the company after leaving the training class. Therefore, in both cases it is extremely important to continuously set goals, check progress, and coach middle managers in developing skills as teachers and facilitators of change. It is useful to develop a behavior profile standard that can be used not as a way to catch people failing but as a way to make the desired behaviors and habits explicit for all (Fig. 9.4).

For this to work, senior leaders must model the desired behaviors and coach others. This can be a great challenge. Committing to personal practice and behavior change must be an early focus of the readiness activity of top management.

Making Time for *Kaizen*

Even after addressing concerns of middle managers, aligning with long-term purpose and short-term business goals, and building enthusiasm for the vision of the future, a very practical question remains: "Where will we find time to do these new things in addition to our regular job?" Transformation will not succeed without engaging everyone in the process,

Change Leader Behavior Profile	Always 2	Sometimes 1	Never 0
Respect for People			
1 Treats people with dignity and respect			
2 Allows people to contribute openly and honestly in the meetings			
3 Encourages people to admit mistakes			
4 Gives constructive feedback on others' performance			
5 Enables people to participate in decisions that affect their work			
6 Listens to people's issues			
7 Gathers relevant information before reaching conclusions			
8 Recognizes the team's positive contributions to the success of the group			
9 Gets people to focus on the issues rather than looking for someone to blame			
10 Ensures meetings start on time			
11 Schedules meetings so that they do not clash with other meetings			
12 Ensures people receive recognition when they do a good job			
13 Ensures that people are rewarded according to job performance			
Continue Learning and Improvement			
14 Ensures that people share ideas and knowledge			
15 Coaches individuals and groups on the performance at their workplace regularly			
16 Teaches subordinates (supervisors, group leaders, managers) to coach and train their people			
17 Makes development plans for subordinates to ensure that all receive the necessary training to do the job			
18 Motivates subordinates to improve their knowledge and skills at work			
19 Asks people to search for and expose problems			
20 Adopts a systems perspective when looking to make improvements			
21 Promotes a team approach to solving problems			
22 Works with people outside of the immediate area to resolve problems			
23 Gives help and support when people require it			
24 Seeks out best practices and promotes these within the group			
25 Facilitates and promotes change			
Process and Results Driven			
26 Involves people in establishing work processes			
27 Ensures that people adhere to established work processes			
28 Responds rapidly when processes go wrong			
29 Ensures that clear procedures are utilized when processes go wrong			
30 Encourages group leaders and work group members to stop the process when quality issues arise			
31 Requires that visual controls alerting work groups to issues and actions are in all areas			
32 Acts on people's suggestion to improve processes in a disciplined way			
33 Ensures decisions are made on a timely basis			
34 Ensures decisions are made on objective rather than subjective information			
35 Checks that decisions are aimed at eliminating the reoccurrence of problems			
36 Holds the correct people accountable for delivering results			
37 Focuses on the customer and the customer's needs, both internal and external			
38 Ensures that people regularly receive the information they need to do the best possible job			
39 Helps people understand the key priorities of the business and how they can contribute towards them			
40 Allows people adequate opportunity to ask questions and raise concerns and deal with them accordingly			
		Total Score	
64–80: Effective, raise to 1, 2s and coach others			
50–63: Effective in many areas, coping in others, raise 0s to 1s and 1s to 2s		0	
49 or lower: Address critical gaps, raise 0s to 1s			

Figure 9.4 Change leader behavior profile.

and this can only be done as a combination of project work and daily work. Often this question comes up because the transformation is presented as a project, a series of trainings, with key performance indicators, reports, and aggressive timelines. Projects may be how big things get done, but learning and practice must happen daily. We have seen that *kaizen* is not something in addition to work, but work itself. It is how we make plans, how we design processes, how we run meetings, and how we solve problems and do things slightly better each time. The cumulative effect is that *kaizen* generates more capacity, but there is an initial investment of time to establish basic systems and routines, learn, and gain competence.

Mark Graban is a consultant who brings his background with *kaizen* in automotive, electronics, and industrial products companies to the healthcare world. He is author of *Lean Hospitals* (2011) and *Healthcare Kaizen* (2012). His work is focused on building strong healthcare organizations that are capable of delivering safe, high-quality, low-cost patient-centered care. Here he explains his response to one of the most common objections raised by leaders when facing the opportunity to introduce a *kaizen* culture:

> I always like to challenge people who say they have no time. We say we don't have time for improvement, which is what leads to the never-ending spiral of things getting worse and having even less and less time for improvement. You have to make the time— so lack of time should be the first problem that you're trying to solve.
>
> Finding time for *kaizen* comes from the willingness of leaders to directly participate in the process themselves. If you're a manager, you need to first identify and implement *kaizen*-style improvement related to your own work and share that with people who are working for you.
>
> What we see in hospitals that are successful is senior-level leaders asking this of their vice presidents, which leads to vice presidents asking their directors, and all the way down the chain. Leaders at all levels have to talk about *kaizen* very frequently, and they have to participate in the *kaizen* process in a very visible way.
>
> Senior leaders need to, in a supportive way, not accept "We don't have time" as an excuse. And by supportive way, I mean don't just beat up on the managers and set a quota for, say, number of *kaizens* per month, because then people start doing lame, trivial *kaizens* to get their senior leader off their back. Senior leaders need to say, "I'm going to work with you to free up time so that you can do kaizen."
>
> With hospitals, I often challenge the practice of sending nurses home early when the patient census is low. They do this to improve short-term labor productivity. But in reality, these same managers who say "We have no time to improve" are throwing away the opportunity.

> People aren't given time to work on improvement because that's not "productive time." Somebody said recently at a conference, "Our daily labor productivity number is the exact thing that interferes with us ever improving labor productivity." These are the constraints or barriers that you have to change.
>
> If we say we want *kaizen*, making time for it has to become part of our leadership culture. Otherwise, "We don't have time" becomes a cheap and easy excuse.

Graban shares at least three key insights here in terms of how leaders should prepare their organizations to implement the practice of *kaizen*:

1. Leaders must understand the cause-and-effect between not making time to fix broken processes and the lack of time to improve because time is lost to these broken processes.
2. Leaders must invest their personal time and energy in grasping both what is possible through *kaizen* and what is the smallest increment of time that could be carved out of a day or week to make meaningful improvements.
3. Leaders must recognize and address incentives, measurements, and policies that unintentionally penalize the proactive problem solving behaviors we wish to promote.

We have the same amount of time in a day or week regardless of whether we spend it on proactive and strategic changes or on reactive firefighting and problem solving. The power of *kaizen* is that it teaches us just how much of our time is lost doing non-value-adding things and can be recovered through small investments in fixing broken processes. The all-important initial investment of time must come from senior leadership.

Finding a Smooth Starting Point

There is a Japanese word *suberidashi* which means "making a start" to a project or a venture. It literally means "beginning to slide" and is an appropriate guide to answer the question, "We think we are ready. Where should we start?" Kotter (1996) and others have pointed out the importance of establishing early successes and quick wins to build belief and confidence in the transformation. Maurer (2004) has argued that smaller changes are better because success builds on success, affecting our brains and minds, increasing our change capability. The last step in organizational readiness is to select pilot areas that are primed for success, where friction is minimized, bumps in the road are removed ahead of time, and the project can "begin to slide," rapidly gaining momentum. While there is no "one size does not fits all" approach to a smooth start, and even within one organization the starting points may differ as projects are launched across a large organization, there are key questions in common that guide the selection of the starting point:

▲ How large of a business impact must we make?
▲ What is the business need?
▲ What will create belief?
▲ What do we need to learn?
▲ What behavior changes and new habits are needed?
▲ How to lead transformation effectively?
▲ How much gas do we have in the tank?
▲ How much time, resource, and management attention is needed?
▲ What is the sense of urgency?

Once the pilot area or areas have been selected, the resources and scope must be planned, taking into account the three areas of daily *kaizen*, project *kaizen*, and support *kaizen* (Fig. 9.5). All transformation efforts must go through a phase when they are project-driven. Many start out as projects to launch a major initiative, projects to build continuous-improvement infrastructures, projects to learn and apply the strategy deployment process (*hoshin kanri*), and so forth. Projects will always be necessary and useful ways to organize large, complex tasks, but they cannot be the only way or even the main way to shape and sustain a *kaizen* culture. However, projects must be directed and steered from somewhere, requiring a steering committee as part of the support *kaizen* structure. Communication, recognition, and motivation will become a permanent part of the new culture and can be tested in the pilot area within the support *kaizen* structure.

The daily *kaizen* must always be part of any project to ensure that the gains made through projects are sustainable beyond the pilot phase, the front-line leaders and managers have a practice field on which to learn new skills and behaviors, and the new span of support models can be tested. In other words, the selection of one or more smooth starting points requires

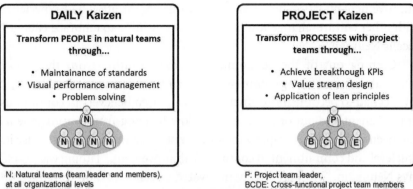

N: Natural teams (team leader and members),
at all organizational levels

P: Project team leader,
BCDE: Cross-functional project team members

1: CEO sponsorship,
T: Top management team sponsorship,
K: Kaizen manager

M: Motivators working for the kaizen manager
Project Management Office / Kaizen Promotion Office

Figure 9.5 The *kaizen* change management structure.

consideration of the desired results and the process that the organization will follow in achieving those results.

Embracing Change at Oregon Community Credit Union

Oregon Community Credit Union (OCCU) is a not-for-profit financial institution based in Eugene, Oregon. Founded in 1956 by a small group of passionate people, the company employs 260 people and is owned today by more than 100,000 members. The company's continued growth and success are based on an unwavering focus on delivering value to members and communities through reinvestment of profits and by providing exceptional personalized service. The interest in *kaizen* came from several places within the organization. Chief Financial Officer Ron Neumann had experience from his career in the manufacturing industry. Chris Whittaker, director of lending services, was an early adopter who had led process-improvement efforts in his area. "Because our organization had experience with other culture-change initiatives, we were willing to embrace change," recalls Neumann.

Coming out of the economic recession resulting from the global financial crisis and facing compression with interest income, it was clear that OCCU had to become more efficient. Having been believers in continuous improvement and out of need, a lot of "low-hanging-fruit" improvements had been picked. The company took this crisis as an opportunity to raise its skill level and commitment to continuous improvement. "Timing was good," says Neumann. "We were highly motivated to generate more revenue and reduce expenses—with a definite focus on improving financial performance. However, we never forget that our mission is to serve our members—so as long as this is at the forefront of our minds as a credit union, it will help us to carry our continuous-improvement efforts long term."

In 2010, the management team began studying *kaizen* and how it might help them get to the next level in member service excellence and cost reduction. In 2011, the senior management team signed on after a workshop on *kaizen*. Management dubbed 2012 the "Year of Knowledge" and began the process of learning by doing, and company-wide goals were set for 2013. The management team developed brief training to explain *kaizen*, 6S, problem solving, and the five whys, both short online presentations, to new employees. *Kaizen* was built into budgeting and thinking through inclusion as one of four key top-management initiatives:

1. *Member relationships.* How many products/services per member on average?
2. *Originating loans.* Targets for new loan originations.
3. *Net income.* Revenue, profit, assets.
4. *Kaizen objectives.* Number of *kaizen* ideas, total involvement.

As spokesperson for entire initiative, Neumann places a high value on two-way communication to keep up the momentum. Directors and vice presidents are updated weekly on the progress of efforts, give feedback before new initiatives are launched, and are asked to promote and support ongoing *kaizen* efforts. Neumann and Whittaker write a monthly one-page message to all employees, giving everyone visibility of performance and progress on projects. Each monthly message also contains a reminder of key words to trigger thinking, such as:

▲ What frustration did you encounter today?

▲ Let's identify and remove pain points in daily activities.

▲ No idea is too small, one penny or one second saved.

From day one, the total workforce has been engaged in *kaizen* in several ways. The first is through 6S activity—the traditional workplace organization method of sort, set in order, sweep, standardize, and sustain, with the addition of security as the sixth S. "We were advised that if you can't pull off 5S, you can't get far with continuous improvement," says Whittaker. Initial emphasis was placed on the auditing process. Only the senior team did auditing at first so that its members could engage directly with the entire workforce in continuous improvement; it also enabled them to become educated enough to understand and lead 6S activities and ensure that they were practical and value-adding. "We made it very clear that 6S is not a clean desk policy; it is about making your everyday work faster, cheaper, and better," says Whittaker.

At all staff meetings, everyone—senior managers, front-line personnel, and facilities staff—reviews metrics and performance, is encouraged to give improvement ideas, and contributions are recognized. "The target for 2013 is to receive 1,500 *kaizen* ideas with everybody's participation. We fell behind early in the year, but with encouragement, we are on track," says Neumann. The Kaizea Board (*kaizen* + idea) was developed as a tool for motivation and communication: *Kaizen* ideas are framed and visible to everyone. "There are 350 ideas on the wall right outside of our largest meeting room and online on the company intranet," says Whittaker. Monthly drawings select a random *kaizen* idea to recognize involvement; the employee who submitted the winning idea is awarded a gift certificate for participation and creativity. "We try to make it clear that everybody benefits from participating and making good strides in becoming more efficient," says Neumann.

"As a credit union, there is a natural intimacy with the customer, the members who are its owners. Whenever we make a change, we ask ourselves, 'How does it benefit the member?'" recalls Neumann. With this in mind, the company was determined that its first full five-day *kaizen* event must have member impact and selected the new-account process because it was taking too long. The *kaizen* team videotaped the interaction between credit union employees and the members, with permission. Reviewing the video for improvement opportunities, the team noticed that the current process required the employee to spend more time with paperwork and on the computer than interacting with the member—an imminent opportunity to improve customer experience. "Making what we do member-centric is in the DNA of OCCU," says Neumann. "*Kaizen* gives us one more way to achieve exceptional service in everything we do."

Although a credit union is a very different business environment than the Toyota Production System, there were distinct cultural values in practice that favored the successful adoption of *kaizen* at OCCU. These include

▲ An atmosphere of respect, decency, and fun in how employees engage with one another

▲ A tradition of being very collaborative within the industry and with other credit unions

▲ Caring about individuals and their health

▲ Giving back to the community

OCCU is beginning to extend *kaizen* beyond their company, reaching out to other organizations within the local community to build a network and exchange information. The *kaizen* effort at OCCU is growing slowly, methodically, gradually—at a sustainable rate and in harmony with cherished organizational values.

Starting with the Idea of Changing Culture

We have explained the critical phase of organizational readiness and the key points for transformation to make a smooth start. In *kaizen*, there is a beginning, but there is no end. In Chapter 10 we will discuss how to apply the plan-do-check-act (PDCA) cycle and *kaizen* process to the hugely important task of facing up to the monster that is our organizational culture.

CHAPTER 10

Facing Up to the Culture Monster

Good character is not formed in a week or a month. It is created little by little, day by day. Protracted and patient effort is needed to develop good character.

—HERACLITUS

Too few leaders who turn to *kaizen* or other business excellence methods looking for performance improvement recognize the importance of addressing culture. Likewise, too few leaders who are working on shaping their culture in positive ways are aware of the necessity of the *kaizen* approach in learning by doing while generating business results. With regard to making changes to organizational culture while improving performance, Schein (2009) advises

Never start with the idea of changing culture. Always start with the issue the organization faces; only when those business issues are clear should you ask yourself whether the culture aids or hinders resolving the issues. Always think initially of the culture as your source of strength. It is the residue of your past successes. Even if some elements of the culture look dysfunctional, remember that they are probably only a few among a large set of others that continue to be strengths. If changes need to be made in how the organization is run, try to build on existing cultural strengths rather than attempting to change those elements that may be weaknesses.

While we agree with many of Schein's observations and ideas, the advice to "Always think initially of the culture as your source of strength" runs counter to our experience. Cultural elements are not only the residue of past successes. Highly dysfunctional, authoritarian, nonadaptive cultures may thrive for years within closed markets or as a result of market conditions, family connections, sheer size, or artificial and unsustainable barriers to entry. Bad cultures can deliver good results by luck or by manipulating the rules. Culture is not always the source of strength, but good character is.

The advice to build on existing cultural strengths is certainly sound. But it is a mistake to assume that past success is an indicator of future success, especially in a rapidly changing world that requires that we adapt constantly. As we face the organization's most urgent issues and identify artifacts that address our problems, we must look deeply within the culture to the assumptions and behaviors that will make the counter-measures succeed or fail. Toward a *kaizen* culture, we must keep and strengthen what is good about our culture and discard elements of our culture that weaken and harm us. Culture eats strategy for breakfast and returns uninvited for lunch and dinner. Within the organizational cultures we have observed, unaddressed cultural weaknesses inevitably sap the cultural strengths.

Although there is wisdom in the argument to "build on your strengths rather than trying to shore up weaknesses," we believe that this holds mostly true for individuals but less so for organizational cultures. By not addressing negative cultural elements or weaknesses, not only does performance suffer, so does morale, motivation, and belief in the organization's direction. This operates at the deepest level of culture, in its basic assumptions and core beliefs, and can turn into a vicious cycle if left unaddressed. As we have shown, even small improvements can empower people and begin to create belief in our ability to continue changing. Organizations are made of people, and developing this competence for people to change is an absolute requirement. If change capability is a cultural weakness, this must be shored up without question (Fig. 10.1).

Kegan and Lahey (2010) argue that leaders who make the conscious decision to grow talent, their people, and themselves are in the best position to succeed. Such growth of talent goes through three mind-sets from socialized, self-authoring, to finally self-transforming. Senior manage-

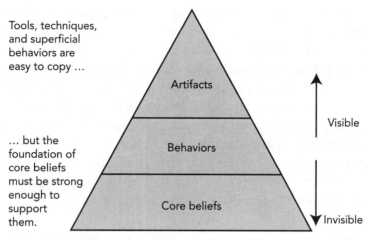

Figure 10.1 Shoring up the foundation of organizational culture.

ment's role is to facilitate individual and collective transformation to reach a higher level of mind-set. What Kegan and Lahey call the "self-transforming mind" is, in many regards, the vision that we have for *kaizen* culture in which the *kaizen* principles are practiced throughout the organization.

But such mind-set transformations do not happen spontaneously. They must be triggered by events that challenge the status quo. Thomas (2008) demonstrated that transformative events called *crucibles* are a cornerstone to force leaders, and therefore their organizations, into "deep self-reflection, where they examine their values, question their assumptions, and hone their judgment," and these crucibles are, in our experience, the events that trigger organizations to transform.

Liker (2003) identified an early crucible moment for Toyota in 1948 when its leader at the time, Kiichiro Toyoda, resigned because of an insolvency problem. That moment marked Toyota's crucible for becoming a fiscally conservative company. We could say that Toyota had a second crucible event in 2010 when quality issues brought its CEO Akio Toyoda to sit before the U.S. Congress and apologize to its customers. Within the company, Toyoda called for a return to basics, a sense of urgency, humility, and reinvestment in people to face the crisis of culture resulting from decades of success at Toyota. In good times or in bad, and even at Toyota, culture is a monster that must be confronted.

A study by Collins (2001) of companies with superior performance over a period of 30 years also identified "confronting the brutal facts" as one of a set of key behaviors that led to success. Facing up to the culture monster requires great humility and courage—humility because each of us makes a place for the culture monster within ourselves, and courage because we know well the power that is our own inaction. An avoidance of performance and culture problems has rarely led to their improvement and certainly not on a sustainable basis. If, then, culture bears so much influence on the success of our enterprises, the growth and happiness of people, and the well-being of society, leaders bear a great moral responsibility to face up to this monster and tame it for good.

Why Transformation Programs Are Failing

We have not failed until we have stopped trying. The act of failing is not the same as the acceptance of failure. Nonfatal mistakes that result in learning are the daily bread of continuous improvement. However, there is a limit to the level of failure that organizations and individuals can endure before the emotional bank account is emptied. When we launch transformation programs, we are placing an unreasonable burden on people to go through major changes, disrupt their routine, and vary their workloads. In the end, transformation programs that place culture at the end or not even within their scope waste much of the effort. When leaders fail to create a cultural environment in which people can practice and keep the skills they learn, the organization is not ready to adopt new habits that are needed for the benefits to be sustained. Over the past three decades, the authors witnessed or were part of transformation programs that were failing, and in each case, it was culture—the assumptions and core beliefs, the behaviors, and visible artifacts that were not in harmony.

A transformation succeeds when we anchor the change in the culture, as Kotter (1996) advises in the eighth step of his eight-step process for leading change. The transformation effort must be *designed and launched* based on values, principles, and cultural elements that will ensure its success. Kotter's steps 1 through 7 cannot happen successfully without a culture that already embraces to an adequate degree the notions of urgency, respect for people, follow-through, consensus, and engagement. The power of the culture monster must be faced, tested, and put to work for the organization.

The *kaizen* approach to change management is to understand the process of how we make decisions, how we communicate, and how we respond to problems and to find the root causes for problems in these areas. This leads us to culture. Transformations will continue to fail until leaders develop the humility to go back to first principles—such as *kaizen*—and understand the true meaning of "change and make good" from the level of core values and basic assumptions. Only by looking deeply into our hearts and minds will we be able to truly answer the questions of whether we are willing to do what it takes to succeed—change ourselves.

This process of changing ourselves occurs largely in three progressive stages: *toolset, systems,* and *mindsets.* We can consider these as analogues to the three levels of culture: artifacts, behaviors, and basic assumptions (Fig. 10.2). Each stage is characterized for mindset plateaus that represent the predominant understanding of *kaizen* within the organization. Stages are built on previous stages once limitations of mind-set become apparent. Most *kaizen* efforts start with the short-term sighting approach of implementing tools in order to quickly reap benefits on improving performance. In our experience, this approach surges out of internal group dynamics in which one group deals with a critical performance problem or searches for wider support from its organization's leadership

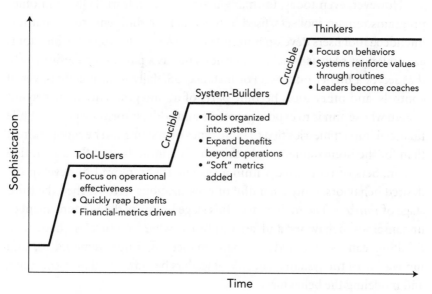

Figure 10.2 Culture transformation from toolsets to mindsets.

(middle-up or bottom-up change). In either case, change is occurring but is not being managed. In other cases, leadership creates the need for change top-down. Gains in operational effectiveness by the use of tools such as single-minute exchange of dies (SMED), mistake-proofing, and flow develop *tools users* within the organization. Financial or "hard" metrics such as return on investment (ROI) are predominantly used to justify deployment at this stage, whereas the development of desired behaviors is yet a limited focus.

Imai (1986) took pains to point out that the *kaizen* programs called total quality control (TQC) and QC activities should not be seen as toolsets aimed at improving product or process quality only. The *Total* was added to QC activities to indicate that TQC aims to improve the capability and performance of management at all levels of the organization. Later, TQC was renamed total quality management (TQM) in recognition of this. In addition, what Japanese managers today mean when they use the terms "QC" or "QC style," is *scientific*. It is understood that QC is not a quality-improvement tool but a systematic, structured, and scientific approach to improvement and that these are desired skills and mind-sets for their people. *Kaizen* builds on these two elements of total engagement of the workforce in a scientific approach to improvement.

However, even today, for many people, *kaizen*, lean, TQM, and other programs remain tool sets, used in ways that are different from what was intended. Tool users develop into *system builders* at the stage in which tools are no longer seen as isolated extensions but as a part of a cohesive whole that serves a higher purpose. For instance, 5S, daily team meetings, visual controls, and other tools become parts of an integral daily management system whose aim is to expose problems and address them as a team. While financial impact metrics tied to these types of systems can be more indirect than for the improvement tools, this is only a limitation of the accounting system. Sets of tools used within a system are employed to develop the desired behaviors. Only a handful of organizations have reached the third stage of *thinkers*. The main aim of this stage becomes the reinforcement of unconscious behaviors and assumptions, where culture is shaped, so thinking can be focused on maintenance of, improvements to, and innovation of the systems. The role of leaders becomes guidance, coaching, and modeling the behavior.

The three stages are similar to the *shu ha ri* tradition of learning of various traditional arts from the masters. Basic learning occurs by copying exactly (*shu*) in order to master the basics, breaking away from the form (*ha*) to adapt them to oneself, and going beyond them (*ri*) to create something different and unique. At the *tools-users* stage, tools are copied as a result of benchmarking higher-stage organizations. As tools are mastered and interconnected to create systems, organizations start the *system-builders* stage. Only when organizations master their system builders can they grow beyond them to improve.

But what about startup and early stage companies that do not yet have set organizational cultures? Startups and early-stage companies have a unique opportunity to get their cultures right and set a critical foundation for long-term success. Kotter and Heskett (1992) identified contributing factors in their study of high-performance organizations. The five most important behaviors in shaping excellent cultures in the successful firms surveyed were

1. Founders clearly articulated mission and purpose.
2. Success built belief in the strategy and mission.
3. Timeless values were explicitly defined.
4. Manager buy-in was gained.
5. Teaching the culture was ongoing.

The first condition can only be achieved at the startup or early stage. Of course, leaders of teams within larger organizations can and do articulate their mission and purpose, but the impact is limited if there is no top-management clarity or linkage. The other four behaviors are part and parcel of a *kaizen* culture. Management is engaged in putting these values into action. *Kaizen* is built on a set of timeless values and principles. The *kaizen* process applies scientific problem solving and creative thinking to build belief through success. If the engagement is weak, the values turn out not to be timeless, and if the results are inadequate, people in a *kaizen* culture do not give up but give the PDCA cycle another turn. While the culture monster may be younger within a startup company, it can be surprisingly strong since startup companies by nature rely on founders with strong viewpoints. Facing up to this is critical to enabling not only survival of the new venture but its development into a lean startup.

Overcoming Cultural Obstacles to Kaizen *at Franciscan St. Francis Health*

Joseph Swartz, Director of Business Transformation at Franciscan St. Francis Health, recalls the cultural challenges that his leadership team had to overcome through their *kaizen* transformation process.

When starting out on the process to engage everyone in *kaizen*, we encountered several cultural obstacles. First we realized that we had a culture of perfection that resulted in an atmosphere of fear of making mistakes. Change happened, but most big changes happened only when it was strongly driven by leadership. Leaders who got things done were rewarded and advanced in the organization. Often it was leaders who were the most demanding and pushed the hardest who were promoted. This ultimately resulted in a culture where employees conform to the leader's demands in order to keep their job. Employee creativity is one of the first casualties in such a culture. We do need strong leaders who can direct the organization firmly when change must happen quickly, but we learned that if we can also lead in a way that engages and empowers our employees, we will reap the rewards of a far more innovative culture.

Second, although it is counter-intuitive, our culture of change management was an obstacle. This manifested itself in corporate committees to steer, control, and manage change. The bureaucratic process became self-serving, rather than serving the purpose of the committees, which was to promote change. A few leaders objected to *kaizen* because they believed it would result in uncontrolled, unmanaged change, which they believed would result in chaos. What they didn't realize at first was that change happens, and many times it happens regardless of how much control we try exert over it. The natural world around us is full of uncontrolled changes, and yet it is part of a larger system of checks and balances. If we want an innovative culture, then what we need to do is to design our continuous improvement system to have unobtrusive checks and balances. One way we do that through our *kaizen* process is to have supervisors coach their direct reports with each *kaizen idea*, and then require their approval on those ideas. It isn't a perfect system because it is dependent on supervisors knowing how to coach well and make good decisions. However, it works well when you have good supervisors.

Third, we learned that simplicity still requires structure. One of our cultural strengths is that we measure ourselves and our performance a lot. At first glance this may seem like waste, against the simple *kaizen* idea, but it was a cultural strength. Applied to the continuous improvement process, we measure the total number of implemented *kaizens*; the dollar savings; participation rates by unit such as department, Vice President, Director, Manager, and other measures. We report a few key metrics regularly to our leadership. Our COO, Keith Jewell, communicated recently that he wants 100 percent of leaders trained in lean six sigma methods and that he wants at least 80 percent of our workforce actively engaged in *kaizen* each year. The ongoing success of our continuous improvement activities is the result of our leaders' willingness to face up to our deeply ingrained cultural beliefs and behaviors, and to adapt these in ways that better served our mission.

Turning the Big Wheel of Culture: Plan-Do-Check-Act

If *kaizen* is the work of making our processes better, creating a *kaizen* culture is the work of getting better at *kaizen*. The true meaning of *kaizen*, and the long-term ideal, is improvement by "everyone, everywhere, every day." Maintenance and improvement of the character and culture of an organization are the ultimate plan-do-check-act (PDCA) cycle of transformational leaders. Managing *kaizen* transformation is an act of leadership that requires a style consistent with the overall values and core beliefs of *kaizen*. Schein (2009) advises that "the best way to assess cultural elements is to bring groups together to talk about their organization in a structured way that leads them to identify their own tacit assumptions. The best way to do this is to first identify all the artifacts pertaining to the area you are inquiring about, especially observed behavioral regularities. Compare these to the espoused values of the organization and, if they don't match, look for the tacit assumption that explains the behavior."

Table 10.1 The Culture PDCA Cycle

PDCA Stage	8-Step Practical Problem Solving Model (TBP)	Culture *Kaizen* Activities
Plan	1. Clarify the problem	Setting clear direction
	2. Break down the problem	Honestly describe the culture
	3. Set the target to achieve	Compelling vision of *kaizen* culture
	4. Analyze root causes	Evaluate *kaizen* core beliefs, gaps
	5. Develop countermeasures	Consensus plan on experiments
Do	6. See countermeasures though	Mobilize people, follow through
Check	7. Check process and results	Look for artifacts and behaviors, verify the underlying thought process
Act/ Adapt	8. Standardize successes, learn from failures and identify gaps for next plan	*Kaizen* the plan, adapting the next turn of the PDCA cycle based on learning from successes and failures

We will not offer a prescriptive method for creating a *kaizen* culture in the reader's organization other than to say, "Embrace the core beliefs of *kaizen*, face up to the culture monster, and turn to the PDCA cycle." We can offer hints on how to adapt the PDCA cycle and encourage the reader to study more about practical problem solving—particularly the eight-step approach modeled directly in what is called *Toyota Business Practice* (TBP). As we can see, great emphasis is placed on clarity of vision, understanding the process, and looking for cause and effect in the plan phase, which enables fast and accurate execution in later phases (Table 10.1).

Furthermore, the teaching of the TBP approach to problem solving is paired with a set of core beliefs called "Drive & Dedication" at Toyota:

1. Put the customer first, both the end customer and internal customer.
2. Always check the purpose of your work, challenging "Why?" from the customer's viewpoint.
3. Take personal responsibility once you have understood the mission and purpose of your work.
4. Visualize the problem for the benefit of others and to promote clarity and alignment.

5. Make decisions based on facts through *genchi genbutsu.*
6. Think the situation through and follow through in action with persistence.
7. Emphasize speed and time-consciousness in responding to changes.
8. Act with sincerity and commitment to doing what's needed.
9. Communicate thoroughly to create alignment towards shared goals.
10. Get everyone involved in getting things done.

During the plan phase, culture *kaizen* activities include setting the direction, aligning with long-term purpose and short-term business needs, communicating a compelling image of the desired future that engages people in the organization, describing good and bad elements of culture and where the *kaizen* spirit is weak or strong, and developing a consensus on specific experiments that will be conducted to test new behaviors.

During the do phase, the leader must mobilize the people; provide effective means to turn the visions into reality at local levels; lead team projects to test the experiments for putting new tools, systems, and behaviors in place; and determine their effect on culture and business results. The leader must follow up with persistence and be visibly involved in supporting the organization to overcome obstacles and challenges that will inevitably appear.

During the check phase of the culture *kaizen* PDCA cycle, it may be necessary to reconfirm or clarify what is negotiable or flexible and what is not with regard to behaviors and how we embody the *kaizen* spirit. Building a culture requires being both inflexible with regard to core values and flexible with regard to how these are turned into artifacts, practices, and visible creations. The imperfect artifact or the misapplication of a tool is not necessarily a sign of people failing to grasp the *kaizen* spirit; it may be a question of competence, which will come with practice. This is why it is important to check the process—the thinking underlying the problem solving—to learn how people are progressing from toolsets to mind-sets. Part of the check phase is to build comfort with the practice of *hansei*—reflection—as a way both to maintain humility and to build confidence and security that exposing mistakes results not in blame but in learning. It is also important to remember that middle-manager resistance may not reveal itself during the denial and anger stages of the change curve we introduced in Chapter 7, but only during the bargaining phase, in the form of asking to be excused from participation, being allowed to skip routines or activities managers see as beneath them, or hanging onto old habits or symbols of power or position. Transformational leaders must be compassionate but firm in guiding these people past the pit of despair and through learning by doing into acceptance of the changes.

In the act/adapt phase, we "*kaizen* the plan" and reflect on what we have learned on the next turn of the PDCA cycle. Take the time to learn, share, comfort, celebrate, find joy, and motivate people to grow and keep developing through *kaizen*. Then find the energy to give the PDCA cycle another turn.

What Problem Are You Solving?

We can say that another definition of organizational culture is its character. As we have seen, what affects character in a person or an organization is a combination of mindsets, moral values, and the integrity with which we act in support of them. Character, reputation, trust—these are attributes that create the "pull" from customers, investors, and talent. Character is a source of competitiveness, but by itself it is not enough. Character must be matched with competence. We have argued that total participation in the practice of *kaizen* enables people and organizations to achieve breakthrough results, develop the competence of people, and embed the values and habits within the culture to sustain the gains. It follows that changing the character and competence of an organization requires changing the character and competence of individuals. This is the true work of leaders. Take the first step, clarify the problem you are solving, and face up to the culture monster.

Bibliography

"Back to the Future at Apple." *BusinessWeek*, May 25, 1998. Available online at: www.businessweek.com/1998/21/b3579165.htm.

Bartlett, Christopher, and Ghoshal, Sumantra (2002). *Building Competitive Advantage Through People*. Cambrdige, MA: MIT Sloan.

Boeing Company (2009). "Lean+ skills benefit community: Sharing our lean+ skills." Available online at: www.boeing.com/companyoffices/aboutus/community/2009_report/lean.html.

Bowen, Kent, and Spear, Stephen (1999). *Decoding the DNA of the Toyota Production System*. Cambridge, MA: Harvard Business Review.

Bucholz, Arden (2001). *Moltke and the German Wars, 1864–1871*. New York: Palgrave Macmillan.

Cerny, Jeff (2009). "10 questions on customer service and 'delivering happiness': An interview with Zappos CEO Tony Hsieh." TechRepublic blog; www.techrepublic.com/blog/10things/10-questions-on-customer-service-and-delivering-happiness-an-interview-with-zappos-ceo-tony-hsieh/1067.

Collins, Jim (2001). *Good to Great: Why Some Companies Make the Leap . . . and Others Don't*. New York: HarperBusiness.

Deming, W. Edwards (2000). *Out of the Crisis*. Cambridge, MA: MIT Press.

D'Mello, Sidney, and Graesser, Art (2012). "Dynamics of affective states during complex learning." *Learning and Instruction* 22(2).

Duhigg, Charles (2012). *The Power of Habit: Why We Do What We Do in Life and Business*. New York: Random House.

Foster, Richard, and Kaplan, Sarah (2001). *Creative Destruction: Why Companies That Are Built to Last Underperform the Market: And How to Successfully Transform Them*. New York: Crown Business.

Geissel, Theodor Seuss (1971). *The Lorax*. New York: Random House.

Grandin, Greg (2010). *Fordlandia: The Rise and Fall of Henry Ford's Forgotten Jungle City*. New York: Picador.

Graupp, Patrick, and Wrona, Robert (2006). *The TWI Workbook: Essential Skills of Supervisors*. New York: Productivity Press.

225

Haidt, Jonathan (2006). *The Happiness Hypothesis: Finding Modern Truth in Ancient Wisdom*. New York: Basic Books.

Hall, Edward T. (1976). *Beyond Culture*. New York: Anchor Books.

Harada, Takehiko (2013). *Mono no nagare wo tsukuru hito: taiichi ohno san ga tsutaetakatta toppu kanrisha no yakuwari*. Tokyo: Nikkan Kogyo Shimbunsha.

Heath, Chip, and Heath, Dan (2010). *Switch: How to Change Things When Change Is Hard*. New York: Crown Business.

Hino, Satoshi (2005). *Inside the Mind of Toyota: Management Principles for Enduring Growth*. New York: Productivity Press.

Huntzinger, Jim (2008). *The Roots of Lean: Training Within Industry: The Origin of Japanese Management and Kaizen*. Available online at: Twisummit.com/wp-content/uploads/2013/02/Roots-of-Lean-TWI1.pdf.

Imai, Masaaki (2012). *Gemba Kaizen: A Commonsense Approach to a Continuous Improvement Strategy*. New York: McGraw-Hill.

Imai, Masaaki (1986). *Kaizen: The Key to Japan's Competitive Success*. New York: McGraw-Hill.

Joho presenter tokudane! Fuji Television Network, Inc., April 20, 2013.

Kahneman, Daniel (2011). *Thinking, Fast and Slow*. New York: Farrar, Straus and Giroux.

Kegan, Robert, and Lahey, Lisa (2010). *Adult Development and Organizational Leadership*. Cambridge, MA: Harvard Business Review.

Kenney, Charles (2010). *Transforming Health Care: Virginia Mason Medical Center's Pursuit of the Perfect Patient Experience*. Boca Raton, FL: CRC Press.

Kotter, John (1996). *Leading Change*. Cambridge, MA: Harvard Business Review Press.

Kotter, John, and Heskett, James (1992). *Corporate Culture and Performance*. New York: Free Press.

Kübler-Ross, Elizabeth (2005). *On Grief and Grieving: Finding the Meaning of Grief Through the Five Stages of Loss*. New York: Simon & Schuster.

Liker, Jeffrey (2003). *The Toyota Way: 14 Management Principles from the World's Greatest Manufacturer*. New York: McGraw-Hill.

Liker, Jeffrey, and Morgan, James (2006). *The Toyota Product Development System: Integrating People, Process and Technology*.

Mann, David (2010). *Creating a Lean Culture: Tools to Sustain Lean Conversions.* New York: Productivity Press.

Maurer, Robert (2004). *One Small Step Can Change Your Life: The Kaizen Way.* New York: Workman Publishing Company.

Meyer, Stephen (1980). "Adapting the Immigrant to the Line: Americanization in the Ford Factory, 1914–1921." *Journal of Social History* 14(1).

Miller, Jon (2003–2013). Multiple articles. Available online at: www.gembapantarei.com.

Myers, Marc (2012). *Why Jazz Happened.* Berkeley: University of California Press.

"Nassim Taleb on antifragility." *Economist,* November 11, 2010. Available online at: www.economist.com/blogs/multimedia/2010/11/nassim _taleb_antifragility.

Nemoto, Masao (1983). *TQC to toppu bukacho no yakuwari: taishitsu kaizen do doukizuke no youtnen.* Tokyo: Nikka Giren.

Ohno, Taiichi (2013). *Taiichi Ohno's Workplace Management.* New York: McGraw-Hill.

Osterwalder, Alexander, and Pigneur, Yves (2010). *Business Model Generation: A Handbook for Visionaries, Game Changers, and Challengers.* Hoboken, NJ: Wiley.

Pascale, Richard, and Sternin, Jerry (2010). *The Power of Positive Deviance: How Unlikely Innovators Solve the World's Toughest Problems.* Cambridge, MA: Harvard Business Review Press.

Pfeffer, Jeffrey (1994). *Competitive Advantage Through People.* Cambridge, MA: Harvard Business Review Press.

Pfeffer, Jeffrey (1998). *The Human Equation: Building Profits by Putting People First.* Cambridge, MA: Harvard Business Review Press.

Porter, Michael (1985). *Competitive Advantage.* New York: Free Press.

Porter, Michael (1996). *What Is Strategy?* Cambridge, MA: Harvard Business Review Press.

Ries, Eric (2011). *The Lean Startup: How Today's Entrepreneurs Use Continuous Innovation to Create Radically Successful Businesses.* New York: Crown Business.

Robinson, Alan, and Stern, Sam (1998). *Corporate Creativity: How Innovation and Improvement Actually Happen.* San Francisco: Berrett-Kohler Publishers.

Rodak, Sabrina (2011). "10 Years Later: Virginia Mason Production System Still Going Strong." *Becker's Hospital Review.* Available online at: www.beckershospitalreview.com/hospital-management-administration/10-years-later-virginia-mason-production-system-still-going-strong.html.

Rother, Mike (2010). *Toyota Kata: Managing People for Improvement, Adaptiveness, and Superior Results.* New York: McGraw-Hill.

Rother, Mike, Shook, John, Womack, James, and Jones, Daniel (1998). *Learning to See: Value Stream Mapping to Add Value and Eliminate Muda.* Cambridge, MA: Lean Enterprises Institute.

Schein, Edgar (2009). *The Corporate Culture Survival Guide.* San Francisco: Jossey-Bass.

Schein, Edgar (2004). *Organizational Culture and Leadership.* San Francisco: Jossey-Bass.

Schmidt, M., and Sommerville, J. (2011). "Fairness expectations and altruistic sharing in 15-month-old human infants." *PLoS ONE.* Available online at: www.plosone.org/article/info%3Adoi%2F10.1371%2Fjournal.pone.0023223.

"The Seeds of Apple's Innovation." *BusinessWeek*, October 11, 2004.

Sekai de ichiban uketai jyugyou. Nippon Television Network Corporation, February 2, 2013.

Shimokawa, Koichi, and Fujimoto, Takahiro (2009). *The Birth of Lean.* Cambridge, MA: Lean Enterprise Institute.

Shirakawa, Shizuka (2006). *Shirakawa shizuka san ni manabu kanji wa tanoshii.* Tokyo: Kyodo News.

Shook, John (2010). "How to Change a Culture: Lessons from NUMMI." *MIT Sloan Management Review.* Available online at: http:// sloanreview.mit.edu/article/how-to-change-a-culture-lessons-from-nummi/.

Taleb, Nassim Nicholas (2012). *Antifragile: Things That Gain from Disorder.* New York: Random House.

Thomas, Robert (2008). *Crucibles of Leadership: How to Learn from Experience to Become a Great Leader.* Cambridge, MA: Harvard Business School Press.

Toffler, Alvin (1984). *Future Shock.* New York: Bantam.

Toffler, Alvin (1991). *Powershift: Knowledge, Wealth, and Violence at the Edge of the 21st Century.* New York: Bantam.

Wakamatsu, Yoshihito (2007). *Toyota shiki sekai wo seishita mondai kaiketsuryoku.* Tokyo: Ryu Books Asti.

Womack, James, and Jones, Daniel (2003). *Lean Thinking: Banish Waste and Create Wealth in Your Corporation.* New York: Free Press.

"You Call That Innovation? Companies Love to Say They Innovate, But the Term Has Begun to Lose Meaning." *Wall Street Journal,* May 23, 2012. Available online at: http://online.wsj.com/article/SB1000142405270 23047917045774182509023099914.html.

INDEX

ABOUT THE KAIZEN INSTITUTE

Founded by Masaaki Imai in 1985, the Kaizen Institute is the pioneer and global leader in promoting the spirit and practice of *kaizen*. Its global team of professionals is dedicated to helping leaders achieve their performance dreams by creating *kaizen* cultures.

The Kaizen Institute guides organizations to achieve higher levels of performance in the global marketplace—easier, faster, better, and cheaper. The Kaizen Institute's *sensei* challenge clients to help develop leaders capable of sustaining continuous improvement in all aspects of their enterprise. The Kaizen Institute creates a worldwide community of practice in *kaizen*.

The major services of the Kaizen Institute include

Consulting and Implementation

▲ Partnering with clients for long-term *kaizen* implementation
▲ Operating-system design and deployment
▲ Breakthrough projects and turnarounds

Education and Training

▲ Business training and academic and online training curriculum design
▲ *Kaizen* practitioner, coach, and manager level certification
▲ On-site training, workshops, and seminars

Tours and Benchmarking

▲ "Kaikaku" benchmark to best-in-class organizations in Japan and worldwide
▲ Building peer-to-peer learning and tour exchange network

Visit www.kaizen.com to learn more about *kaizen* and the world-changing purpose of the Kaizen Institute.

WORLDWIDE CONTACT INFORMATION FOR THE KAIZEN INSTITUTE CONSULTING GROUP

―――――

Americas

United States

7137 East Rancho Vista Drive, B-11
Scottsdale, AZ 85251 USA
Tel: +1 480 320 3476
Fax: +1 480 320 3479
Email: usa@kaizen.com

México

Av. Chapultepec 408
Int. 3 Colinas del Parque
78260 México
Tel: +52 444 1518585
Email: mx@kaizen.com

Brazil

Al. dos Jurupis
452 – Torre A – 2º
Andar 04088-001
São Paulo – SP Brazil
Tel: +55 (11) 5052 6681
Fax: +55 (11) 5052 6681
Email: br@kaizen.com

Chile

Av. Providencia 1998 of. 203
Providencia, Santiago, Chile
Tel: +52 (0) 2 231 1450
Email: cl@kaizen.com

Asia Pacific

Japan

MetLife Kabuto-cho Bldg, 3F
5-1 Nihonbashi Kabuto-cho
Chuo-ku, Tokyo 103-0026
Japan
Tel: +81 3 5847 8245
Fax: +81 3 5847 7901
Email: jp@kaizen.com

China

1027 Chang Ning Road,
   Suite 2206
Shanghai, China
Tel: +86 (0) 21 6248 2365
Email: cn@kaizen.com

Singapore

20 Cecil Street
#14-01 Equity Plaza
Singapore 049705
Tel: +65 (0) 6305 2410
Email: sg@kaizen.com

New Zealand

15a Vestey Drive, Mt Wellington
Auckland 1060 New Zealand
Tel: +64 (09) 588 5184
Email: nz@kaizen.com

India

Office No. 1A, Second Floor
Sunshree Woods Commercial
   Complex, NIBM Road
Kondhwa 411 048 Pune, India
Tel: +91 92255 27911
Email: in@kaizen.com

Europe, Middle East, and Africa

Germany

Werner-Reimers-Strasse 2-4
D-61352 Bad Homburg
Germany
Tel: +49 (0) 6172 888 55 0
Fax: +49 (0) 6172 888 55 55
Email: de@kaizen.com

France

Techn'Hom 3
15 Rue Sophie Germain
F-90000 Belfort, France
Tel: +33 145356644
Fax: +33 145356564
Email: fr@kaizen.com

United Kingdom

Regus House
Herald Way
Pegasus Business Park
Castle Donington DE74 2TZ
UK
Tel: +44 (0) 1332 6381 14
Email: uk@kaizen.com

Netherlands

Bruistensingel 208
5232 AD, 's-Hertogenbosch
Netherlands
Tel: +31 (0)73 700 3440
Email: nl@kaizen.com

Spain

Ribera del Loira, 46 Edificio 2
28042 Madrid, Spain
Tel: +34 91 503 00 19
Fax: +34 91 503 00 99
Email: es@kaizen.com

Saudi Arabia

Al Tawoon Center
Al Tawoon District
Othman Bin Affan Road
Exit 7
Riyadh, Kingdom of Saudi Arabia
Tel: +966 (0)11 21 40 756
Email: sa@kaizen.com

Switzerland

Bahnhofplatz
Zug 6300 Switzerland
Tel: +41 (0) 41 725 42 80
Fax: +41 (0) 41 725 42 89
Email: ch@kaizen.com

Portugal

Rua Manuel Alves Moreira, 207
4405-520 V.N.Gaia, Portugal
Tel: +351 22 372 2886
Fax: +351 22 372 2887
Email: pt@kaizen.com

Italy

Piazza dell'Unità, 12 40128
Bologna, Italy
Tel +39 051 587 67 44
Fax: +39 051 587 67 73
Email: italy@kaizen-institute.it

Kenya

c/o KAM – Kenya Association of
    Manufacturers
3 Mwanzi Road
Opp Nakumatt Westgate
Westlands
Nairobi, Kenya
Tel: +254722201368
Email: afr@kaizen.com

Additional Kaizen Institute Locations

Austria, Belgium, Canada, Czech
    Republic, Hungary, Malaysia,
    Poland, Romania, Russia

To find contact information for
all locations, please visit
www.kaizen.com.

Gemba Academy Online Training

www.GembaAcademy.com

Kaizen Institute Blog

www.gembapantarei.com

Executive Master's Degree Program

www.kaizen.com/Master